Racisms

Renato Foschi

Racisms

The Hidden Logics of Prejudice

palgrave
macmillan

Renato Foschi
Department of Dynamic and Clinical Psychology, and Health Studies
Sapienza University of Rome
Rome, Italy

ISBN 978-3-032-22961-8 ISBN 978-3-032-22962-5 (eBook)
https://doi.org/10.1007/978-3-032-22962-5

Previously Published by Mondadori Università

Cover credit: RyanJLane

This Palgrave Macmillan imprint is published by the registered company Springer Nature Switzerland AG
The registered company address is: Gewerbestrasse 11, 6330 Cham, Switzerland

Foreword

Long ago and far away in 1980, I found myself sitting alone in a train compartment in Italy, accompanied only by a respectable-looking elderly lady. We started talking, and I shared a little about myself. I told her that my father was Italian and from Salerno. "Ah," said the kind lady, "a *terrone*!" (Terrone is a derogatory term used by Northern Italians to describe Southern Italians, and essentially means *ignorant peasants*.) At that time, I was living in England and, having never lived in Italy, I was only vaguely familiar with the Italian North-South prejudice. This comment from an apparently polite and proper lady came as somewhat of a surprise. My mother was Dutch, and, from my experience in England, I knew well that prejudices against Italians in England or in the Netherlands did not make subtle distinctions about regional Italian origins. The northern roots of this "friendly" lady would not have impressed anyone in Northern Europe.

Renato Foschi begins this brilliant and concise guide to racisms with a chapter entitled *We are Always Somebody's Southerner*. If this book had been at my disposal during that train ride, this is how I would have responded to my hostile interlocutor: *We are always somebody's southerner, dear Madam*.

Foschi guides the reader through the many ways in which racism has manifested, focusing, among other things, on how scientific research has been used and abused in the process. The arbitrariness of racism and how science has been co-opted to transform implicit prejudices into explicit categorizations and legislation emerge very clearly in his narrative.

Returning to my opening story, in the United States, where I live, Italians in New Orleans on March 14, 1891 were the victims of a horrific mass murder by an out-of-control yet politically sanctioned mob. This event has been described as the largest mass lynching in American history. While this claim is debatable, the fact remains significant for the transformation of the status of Italians, particularly those from the South. Considered "fearsome" and "murderers by nature," they were regarded as "African" in all respects, partly because they accepted "black" labor in Louisiana's sugar fields and lived among and fraternized with African Americans (Staples, 2019). Jennifer Guglielmo (2012) summarizes the perspective that

> Southern Italy was more than a geographical space (…), it was a metaphor for anarchy, rebellion, poverty, and lack of "civilization." (p. 9)

She goes on to state that prejudice against Italians was often virulent:

> Italians also learned that they were racially "other" in the United States in ways that went beyond language: lynchings; the refusal of some native-born Americans to ride streetcars with or live alongside 'lousy dagoes[1]'; the exclusion of Italian children from certain schools and movie theaters, and their parents from social groups and labor unions; segregated seating in some churches; and the barrage of popular magazines, books, movies, and newspapers that bombarded Americans with images of Italians as racially suspect. (p. 11)

Guglielmo adds this important quote from historian Matthew Jacobson: "It was not just that Italians did not look white to certain social arbiters, but that they did not *act* white" (p. 11). Nevertheless, Italians, even those from the South, ultimately became "white." How this occurred is too complex to summarize here, but suffice it to say that the founding figure of Christopher Columbus played an important role. It should be noted that by 2022, Columbus's status had changed significantly. He entered

[1] Grave insult of uncertain etymology used in the United States against Italians and Latin populations.

the pantheon of historical figures to be erased, with "Columbus Day," a national holiday, now gradually replaced by Indigenous Peoples' Day.

In the United States, the "one-drop rule," which was widespread in the South and held that anyone with even a single "black" ancestor is therefore "black," remains a criterion in the minds of many people. In Brazil, however, there are five official classifications of race and skin color: White [Brancos], Brown or mixed [Pardos], Black [Pretos], Yellow [Amarelos], and Indigenous [Caboclos]. These classifications are unpopular. Brazilians have used many other categories to describe themselves, including "morenos," "negro," and "afrodescendente," as well as numerous humorous colors used by those who reject categories entirely, such as *cor-de-burro-quando-foge*, or the color of a donkey when it runs away.

Questions arise regarding how official categorizations are made, by whom, whether science is involved, and what enormous damage they have caused over the years. Who decides if a person is "black" or "white"? Increasingly, younger Western generations self-identify racially, rejecting old and seemingly arbitrary classifications. An extreme and problematic case in the United States has been Rachel Dolezal, leader of the Spokane, Washington chapter of the National Association for the Advancement of Colored People (NAACP), who claimed to be "black" but was actually "white." Dolezal was heavily criticized as a fraud, but this rather bizarre story illustrates the complexity of the issue and its consequences (Holland, 2015).

Foschi leads us into the scientific quagmire of eugenics and the madness of these fevered Frankenstein dreams of humanity's control over a barely understood nature. In a profoundly mistaken effort of "race improvement," enormous arrogance was accompanied by confidence in science's ability to determine what was best for society, "the race," and "progress."

The picture of racism that emerges from this work is complex. It spans various perspectives, from human psychology to ideologically motivated scientific research, to how politicians use prejudice and manipulation to unite and distract from urgent social issues they cannot solve. During an economic recession or depression, there is nothing better than a racial enemy to unite people and make them forget their problems (Crisp, 2015). The population can be redirected from complex economic and political issues to a simple "enemy," an "other" who can be blamed for all their problems.

In my personal experience, I clearly saw this dynamic in England in the 1970s. Most of Europe was in trouble, and Italy was obviously in its "Years of Lead." After swinging London and the 1966 World Cup football victory, the situation in England rapidly declined. Strikes, blackouts, shortages, unemployment, an oil embargo, and a general moral malaise overwhelmed the country. The situation was complicated. The empire was gone, the global economy was in shock, and starting in 1973, the Irish Republican Army began bombing London. Political leadership was already faltering: on the eve of the 1970 elections, which he unsurprisingly lost, Labour leader Harold Wilson made an inspired remark: "*As much as people may be tired of me, I believe most people will consider me the lesser of two evils.*"

It was therefore no surprise to see explicitly racist and neo-Nazi organizations such as the British Movement and the National Front draw thousands of disaffected, chronically unemployed, hopeless young people with no sense of community or identity. "There's no future in England's dreaming," sang the Sex Pistols, and the words resonated. These organizations gave these young people a sense of identity, community, a mission, and, most importantly, a simple target in an era of confusion: someone to blame.

Even members of the Commonwealth from Jamaica, India, and Pakistan, who after the war had been invited to come work in Great Britain—which was said to be their "home," their "motherland"—were now discriminated against and often violently attacked. Racism has always been actively promoted for political reasons, exploiting humanity's worst instincts during times of crisis and despair.

Foschi leads us through an interconnected web of prejudices that includes antisemitism and misogyny. We recall Frantz Fanon's observation that

> At first glance it might seem strange that the anti-Semite can be equated with that of the negrophobe. It was my philosophy teacher from the Antilles who reminded me one day: "When you hear someone insulting the Jews, pay attention; he is talking about you." And I believed at the time he was universally right, meaning that I was responsible in my body and soul for the fate reserved for my brother. Since then, I have understood that what he meant quite simply was that the anti-Semite was inevitably negrophobe. (Fanon, 1952/2008, p. 101)

In his discussion of sexism, Foschi returns to an important theme: what is or is not considered "natural," particularly the supposed "natural" roles of women and men. It is unsurprising that with the emergence of explicitly authoritarian politicians worldwide, there is increasing polarization of gender roles purported to be "natural." The LGBTQ+ movement and women's political and economic equality are rejected as unnatural. In Brazil, President Bolsonaro's hypermasculine demeanor and rejection of LGBTQ+ and Indigenous communities were central to his image and policies. In China, Xi Jinping criticized "boy bands" for not conforming to the Chinese image of masculinity. Vladimir Putin's shirtless horseback photo reinforces his macho image, as do, apparently, his transphobic statements and a "gay propaganda law." In the United States, the Supreme Court has abolished women's constitutional right to abortion and protections for LGBTQ+ and women are being rolled back.

Again, we face a complex phenomenon. In the United States, for example, women attending university outnumber men and are increasingly the main breadwinner in couples. For some, this disrupts the so-called natural order of society and family. Framing the issue in terms of natural order (almost always willed by God), with the promise of a return to "natural" order, may attract people experiencing a sense of "future shock," a term futurist Alvin Toffler used to describe the confusion and disorientation that occur when people are overwhelmed by loss of a sense of control because everything—even the most "ordinary" things—are now so different (Toffler, 1984).

Foschi does not hesitate to emphasize how spiritual, religious, and esoteric traditions have influenced cultural interpretations of what is or is not part of the "natural order." The popularity of those who may be described as right-wing esoteric thinkers, such as Julius Evola and Aleksandr Dugin, among political movements in Europe and America is worrying but symptomatic of a need for a sense of direction beyond the merely economic.

What can counter the ongoing presence of racism in our world? Education remains a vital force because it is clear that the kind of thoughtful reflection on racism presented in this work has not yet reached the point of being assigned reading for every student in every school. The level of discourse is still too uninformed, and manipulating people with the classic authoritarian playbook—finding someone to blame who does not look like "us"—remains all too easy. I hope this volume is read widely—very widely—and informs the discourse.

Foschi concludes this excellent book with a beautifully poetic passage taken from Pasolini's film *What Are Clouds?*:

> "Ah! Heartbreaking, wonderful beauty of creation!"
> Othello and Iago are united by their wonder at the beauty of creation.

I cannot conclude without returning to my experience in London, witnessing young men drawn into the hysteria and violence of right-wing movements, because it was not the only thing that happened during that period. The second half of the 1970s in London also saw an explosion of musical creativity. Bands such as The Police, Dire Straits, The Clash, Elvis Costello, Nick Lowe, Steel Pulse, and many others played in London pubs and clubs and then traveled worldwide. Particularly relevant for our purposes was the Two Tone movement, featuring bands like The Selecter, The Beat, and The Specials. Two Tone was the name of the record label, but also reflected the bi-racial groups of (Two Tone) black and white musicians playing together before mixed, ecstatic audiences who rejoiced that they were together despite all their struggles. Their union symbolized the society they desired, dancing, singing, and playing music, providing a joyful alternative to racism. Perhaps, even if only for a moment, they experienced and were grateful for the heartbreaking, incredible beauty of human creativity.

September 2025

Alfonso Montuori
California Institute of Integral Studies
San Francisco, USA

References

Crisp, R. J. (2015). *The social brain: How diversity made the modern mind*. Robinson.

Fanon, F. (2008). *Black skin, white masks*. Grove Press. (Original work published 1952)

Guglielmo, J. (2012). Introduction: White lies, dark truths. In J. Guglielmo & S. Salerno (Eds.), *Are Italians white? How race is made in America* (pp. 1–6). Routledge.

Holland, J. J. (2015). Choosing your own: Definition of race becoming fluid. *The Seattle Times*. https://www.seattletimes.com/nation-world/choosing-your-own-definition-of-race-becoming-fluid/

Staples, B. (2019). How Italians Became 'White'. *The New York Times*. https://www.nytimes.com/interactive/2019/10/12/opinion/columbus-day-italian-american-racism.html

Toffler, A. (1984). *Future shock*. Bantam.

Competing Interests The author has no competing interests to declare that are relevant to the content of this manuscript.

Contents

About the Author

Renato Foschi is Professor of History of Psychology and History of Science at Sapienza University of Rome, Italy. He has published extensively on the history of psychological knowledge, including studies on twentieth-century psychological traditions, Maria Montessori's pedagogy, the history of psychotherapies, and the relationship between politics and psychology.

List of Figures

CHAPTER 1

We Are Always Somebody's Southerner: Introduction to a History of Racisms

Traditionally, the term racism refers to an ideology according to which there exist human groups called races, distinct and arranged on a hierarchy in which one is considered better and more desirable than others. The criterion of skin color has been used the longest for such classification and hierarchization. Jackson and Weidman highlighted the distinguishing features of the racist ideology:

1. People can be classified into distinct, biological groups on the basis of physical characteristics, either phenotypic or genotypic.
2. These groups can be ranked on a hierarchy with some superior to others.
3. Outer characteristics are linked to inner characteristics. In other words, some physical characteristics, such as skin color, are a sign of inner, unobservable attributes, such as intelligence, temperament, or moral capacity.
4. These outer signs and inner capacities are inherited and innate. You are born into your race, and you are always the same race as your parents.
5. These differentiated races are fixed either by nature or God. You cannot escape your racial classification (Jackson & Weidman, 2006, pp. XIII–XIV).

R. Foschi, *Racisms*, https://doi.org/10.1007/978-3-032-22962-5_1

These racial ideas have permeated science and philosophy. More specifically, racist scientists have justified their ideas based on biological motivations. However, from a scientific standpoint, ranking human races is a practice so fraught with fallacies that scientists have never agreed on the existence of races (the number and types have varied continuously); the logical consequence is that the concept of race for humans is therefore nonsensical and not objective. Essentially, racist scientists have never succeeded in classifying humans into races, and this will never be possible because humans have never developed biologically distinguishable races; the phenotypic differences observed (skin color and somatic traits) are not as relevant as racists claim for the purpose of distinguishing human races; a classification of races can instead be done for some animals artificially selected by humans.

Despite apparent differences, humans are genetically homogeneous, extremely adaptable, increasingly biodiverse, but not distinguishable into races as homogeneous sets of some kind. Certain animals, when segregated by humans in stable environments with specifically chosen sexual crossings to bring out certain physical types, have exhibited biologically significant genetic differences sufficient to create distinct races.

Humans, on the other hand, are organisms that throughout history have never been segregated enough to produce "racialized" genetic differences. It is now widely demonstrated that humans all derive from a relatively recent African origin; our species is characterized by excellent homogeneity and continuous migrations whereby various parts of the world have been colonized.[1] Superficial physical differences would be the product of recent genetic evolution with a differential value insufficient to determine racial differences; in fact, continuous mixing and migrations will likely increase in the future, producing individual biodiversity that can never be reduced to subdivision into races.

The need for books like this arises from the significant problem that, despite the scientific impossibility of describing human races, there has been a long cultural and scientific tradition that axiomatically presupposed the existence of human races; physical criteria of various kinds were used and correlated with psychological, historical, religious, or more loosely cultural differences believed to be determined by the biological essence

[1] Genetic advances reveal that ancient human migrations likely facilitated interbreeding among different hominids; ultimately, a single genetic root originating from Africa prevailed (Cf. Barbujani, 2016).

of a particular race. Such a general orientation gave rise to prejudices that easily led to various discriminatory attitudes and behaviors. Given that empirically racial distinctions have proven unattainable, the use of the race category applied to humans remains widespread despite studies and legislation attempting to diminish it (Saini, 2019).

This volume does not dwell too much on the scientific irrationality of the biological race concept—for that, one may refer to the excellent works of Guido Barbujani (2005, 2006; Barbujani & Cheli, 2022)—but, assuming the non-existence of human races, it intends to historicize the different forms of racism, expanding the racism discourse to discriminatory profiles. This logic, integrating the meaning of racism, was perfectly present, moreover, in the theorists of racism. For example, browsing the pages of Julius Evola (1898–1974),[2] frequently encountered later as a major propagandist of racism and prejudice, makes clear how his reasoning—which has many followers—is perfectly applicable to sexist, homophobic, or generally intolerant biases (Cassata, 2003; Rota, 2003).

Evola, often encountered as an advocate of both spiritual/cultural and scientific racism (though attempting to distance himself from the latter), defined racism as the basis of a sort of ultraconservative revolutionary ideology, a mentality that is never neutral[3]:

anti-egalitarian
anti-universalist
anti-individualist and therefore totalitarian
anti-rationalist
anti-evolutionist.

From the breadth of these points, it follows that racism is not a matter confined to history or biological research but instead involves culture and politics more broadly. Contemporary authors and historians of esotericism, such as Gary Lachman, have pointed out, among other things,

[2] Evola, who is cited multiple times, was a *sui generis* intellectual, a right-wing theorist, connoisseur of Eastern philosophies, scholar of sexual magic and esotericism, and a traditionalist. He began as a Dadaist and became a traditionalist "philosopher," always remaining racist.

[3] A thorough analysis of the racist conception as the ideological base of Fascist and Evolian traditionalist politics, with extensive bibliographic references, is found in chapter 8 of Cassata (2003).

a hidden influence of Evolian racist and traditionalist ideologies on contemporary international political ideologies.[4]

Moreover, racism has mostly been treated in the singular. However, history clearly shows that it is a composite phenomenon, ranging from the mildest form of prejudice in the average person to extreme forms, such as the Nazi idea, where "undesirables" were to be eliminated by the state based on an artificial concept of evolution. Racism could therefore concern elaborate pseudoscientific theories, typical especially of the positivist and post-positivist periods, as well as the common sense of ordinary people who daily judge others based on unfounded prejudices about human groups, built on intuition and appearances without a logical or experimental basis.

Methodologically, it must be stated that contemporary scholar Pierre-André Taguieff (1988/1994, 2011), a noted theorist against racism who argues it would be an error to extend the term racism to all prejudices, is not entirely convincing because racism has never been a unique phenomenon; moreover, the racist mentality always uses the general logics of discrimination (which was explicit even in self-declared racists). This is all the more true because anti-racist activists have, from the beginning, collaborated in typically transversal anti-sexist and anti-homophobic struggles. Racism, therefore, far from being exclusively a notion linked to "races" is a much more complex and certainly plural phenomenon.

Racism, therefore, is not a unified concept, but rather a series of prejudices with different nuances and justifications concerning the idea of hierarchy and inequality among human beings. According to La Vergata (2009, p. 134):

> Racism has had many forms and representations. There were: paternalistic and violent racism; environmentalist and hereditarian; mystical and scientific (that is founded on what, in a specific historical-cultural context, appear as scientific data); one according to which inferior races could have succeeded but did not; one according to which they could succeed if guided by superior races; one according to which they never could; one for which they must be controlled and one that believes they should be

[4] Lachman notably helped popularize the history of esotericism in the Anglo-Saxon world, distinguishing between progressive esotericism and radical right-wing esotericism. See his writings, especially Lachman (2018).

enslaved or treated without scruples; interventionist and laissez-faire; segregationist in the name of racial purity and one in favor of racial mixing (or at least some of them); more biological and more historical-cultural (with all intermediate varieties possible); transnational elite racism (a racism of the spirit) and nationalist; pessimistic and catastrophic and optimistic and even utopian; …

According to science philosopher Ian Hacking, it has been chiefly an instrumental and tribal concept for differentiating between an "us" and a "them," between friends and enemies. The former were considered bearers of desirable values, and enemies were characterized by values to reject. Such ideas about enemies were primarily unfounded prejudices. The tribal and instrumental roots of the concept have enabled it to endure for such a long time (Hacking, 2005). In its most subtle and perverse aspects, the racist tribal attitude is even unconscious. In a war, one is outraged for victims of the same race, religion, or culture close to theirs, while victims perceived as different cause less indignation and less willingness to analyze and propagate the facts. The distinction between "us" and "them" has been extensively studied by psychologists (cf. Guan & So, 2023).

Social psychology has developed theories based on experimental experiences, founded on concepts such as Social Dominance Theory (SDT), which posits that some human groups control and dominate others (Sidanius et al., 1992). In social dominance theory, the elitist and negative logics that lead to the discrimination of one power group over another are analyzed and demonstrated. At the base of these logics are cognitive and emotional mechanisms that favor splitting between good and evil, right and wrong, black and white. Mental mechanisms studied in psychology and psychoanalysis include similar concepts, such as authoritarianism, moral absolutism, totalitarian mindset, mental polarization, splitting, and the clinically recognized concept of schizoparanoid mentality, as elaborated by Melanie Klein (1882–1960) (Lauriola et al., 2015. Cf. Montuori 2005).

In the epistemological path to define the use of racialization, we follow Hacking. John Stuart Mill, in his treatise *A System of Logic* (1843), devoted an exposition to racial classifications that became a starting point for Hacking. Human differences, as Mill saw them, could be considered self-evident, as well as the fact that some human groups exhibited characteristics in common that others did not. Mill—who was not racist—considered differences between humans as superficial elements. Logically, to determine the existence of human races, strong differentiating elements other than skin color, which is superficially determined by evolutionary processes, would be required. Despite scientific efforts, many concretely differentiating and varied elements have never been found, and, as Hacking notes, will likely never be found (Hacking, 2005, p. 142). The category race, therefore, has an instrumental purpose, which can be, for example, to maintain slavery or serve a cultural or political purpose. It is based on differentiating elements that, even if natural (such as a predisposition to specific pathologies), are not constitutive of human nature.

Despite the superficiality of differentiating elements, the efforts by racists to prove the existence of human races and their supposed hierarchy represent a good didactic example of how differential research categorizes people into sets, looking for criteria that seem concrete but are not. A racist conception indeed needs to define different races, quantify and correlate elements within the same category, normalize human categories, find biological and genetic foundations for categorization, and plan administrative tools to manage said categories. Finally, categorized people attempt to reclaim the categories in which they are classified.

Initially, geographical origin, body morphology, and skin color were the factors that defined the different racial categories used to describe humans. These racial classifications entered medical use with the Frenchman François Bernier (1620–1688) in the late seventeenth century. They were subsequently useful for regulations by Louis XIV on the slave trade, where enslaved people were considered to belong to an inferior race (West, 1982).

The classifications of Linnaeus (1707–1778) and creationism then provided the cultural context that helped the idea of populations born in different places with stable and immutable characteristics at birth. With Linnaeus, humans were divided into four categories: red (Americans), yellow (Asians), black (Africans), and white (Europeans). Morphological and psychological characteristics were considered emanations of divine creation, which, however, due to these religious beliefs, implied a hierarchy among human groups, with some considered superior and others inferior. Such classificatory and creationist tendencies persisted for a long time, at least until French evolutionism began to challenge creationist and fixist tendencies, particularly with Buffon (1707–1788) and his student Lamarck (1744–1829) (Jackson & Weidman, 2006, pp. 15–18).

Doubts arose about the immutability of races and their creation by God. However, evolutionism also marked the beginning of a different kind of racism from the religious one, convincingly supported by scientists starting with Buffon, who interpreted biological data to confirm racial prejudices, including those of religious believers (Curran, 2012). Within science, a sort of confrontation between racists and anti-racists started and continues to this day. Buffon, for example, while critical of fixism and simplifications by Linnaeus, was convinced of the superiority of white Europeans and believed that other human types were the result of the degeneration of the white race due to hostile environmental factors. For these predecessors of ours, such reasoning justified by religion or science was therefore taken for granted.

Physical anthropology, which measured bodies and skulls, provided an empirical basis for racism; this discipline, very popular until at least the Second World War, correlated physical elements to describe racial categories. The subsequent step was genetics, used to select apparent races and the most desirable individuals belonging to these pseudo-races through eugenics. Racism was then also the fear of miscegenation or safeguarding presumed racial purity. Populations in the Global South, especially those of African descent, were considered to have different somatic and evolutionary characteristics, with features perceived as regressed compared to those of northern peoples.

These racist ideas were also rooted in an interpretation of the Old Testament, whereby the curse of Noah on his grandson Canaan, son of Ham, guilty of offending his father Noah, legitimized by God's will the fact that all populations descending from Abrahamic religions (Jews, Christians, Muslims) could enslave Africans as the lineage of Ham and Canaan, both cursed by Noah (while other populations, European and Asian, were thought to descend from Noah's other sons) (Jackson & Weidman, 2006, pp. 4–6). Racisms justified by biblical interpretations thus flourished due to a political culture mixed with religion, which used these prejudices to justify the domination of one people over another, administer conquered territories, and manage real "empires." The interpretive line of populations as descended from Noah's sons led to the concept of so-called polygenism, or the "scientific" belief in the evolution of races from various progenitor types, as opposed to monogenism, which held that evolution from a single progenitor was more accurate. The first polygenist racist anthropologists sought various lineages in skull differences; others remained monogenists, but also based on skull conformity considerations. Both polygenists and monogenists, following different logics, could be racists.

Hacking (2005) noted that despite the biological concept of race having been repeatedly declared unfounded by important institutions and scientists, it persists for multiple reasons, even within epistemologically different frameworks other than traditional ones. Among the reasons for its persistence are cognitive logics that categorize the world into friends and enemies, as well as new scientific disciplines that study susceptibility to diseases in specific human groups sharing attributes that, although superficial, predispose them to one malady rather than another. For example, northern peoples are susceptible to skin cancers because their skin is more exposed to sun damage. Such studies in genetic medicine, which have little to do with racism proper, can nonetheless conceal beliefs by contemporary researchers very similar to those of anthropologists and biologists of the past.[5] Thus, even racism in its pseudoscientific foundations may cyclically resurface under new guises seemingly distant from old racism, but which in reality can become a presupposition, now as then (Barbujani, 2006).

In the eighteenth century, during the Enlightenment, contradictions began to emerge in racial differentiations as witnessed by Mill's considerations, which over time led to an emancipation process of subordinate human groups and a fight against racism that gradually meant the reclaiming of identity by those classified and hierarchized. The phase of reclaiming one's identity, also analyzed by Hacking, played a significant role in history, as it shaped contemporary cultures of pride and claims of rights among those who had been racialized or persecuted due to prejudice.

The Enlightenment promoted a culture of equality, opposing racism, and humans began to view equality with different attitudes. The motto *Si tu tiens aux distinctions humaines, Sors. Nous n'en connaissons point ici* (If you insist on human distinctions, leave. We know none here) was one of the principal mottos for one initiated into French Freemasonry of the eighteenth century. This egalitarian culture became a key element in advanced societies, slowly entering political thought, constitutional charters, and human rights declarations stemming from liberal revolutions.

[5] This is the main theme developed in Saini (2019).

Race discrimination among humans was a concept that clashed with the advanced ideas of the eighteenth century. However, conservative ideas had not yet been supplanted, and the political-economic interests of colonial empires predominated over democratic and Masonic ideals. Some political figures prioritized nationalist interests over universal ones. Colonial imperialism for centuries justified racism. Colonizers thought themselves right, assimilating entire populations considered inferior, non-Christian, and primitive. Many examples exist from Spanish conquests eliminating entire South American cultures after Columbus to Italian fascist racism attempting to build an empire annexing Ethiopia, passing through English and French imperialism, and the genocide of Native Americans by the United States. It is evident that Enlightenment ideas were also divisive, and values of equality, fraternity, and liberty were interpreted differently to justify politically unjustified actions. There were universalist interpretations promoting abolition and respect for humanity and particularist ones applying these values only to whites or gradually to "colonized" such as African Americans, especially in the United States. These considerations were particularly valid for the United States, a country founded on such democratic ideals interpreted in polarized ways; the Civil War saw many soldiers linked to opposite Masonic lodges fighting for ideals interpreted differently.[6]

Between the eighteenth and late nineteenth centuries, slavery was finally abolished first in Europe and then in the United States, from northern states to southern states, culminating in the Civil War. The final abolition victory did not lead to a decline in racism, which paradoxically flourished as a white reaction to abolition, becoming essentially segregation (Jackson & Weidman, 2006, pp. 33–34).

As ideals of liberty, equality, and fraternity spread, an inclusive and antiracist "stop and go" process began, evolving over decades from a trickle into a river. Starting with somewhat abstract and contradictory petitions typical of advanced Enlightenment circles, egalitarian culture nonetheless began to penetrate, addressing increasingly concrete aspects of society.

[6] Recently, for example, a time capsule was found at the base of the statue of General Robert E. Lee in Richmond, removed in 2021, containing two objects in a single envelope: a Confederate flag and a Masonic medallion. Both Northern abolitionists, many of whom were Masons, and Southern racists embraced similar values, yet interpreted them so differently that it led to war between "brothers."

Generally, the nineteenth- and twentieth-century human rights movement, which included labor rights, women's rights, African Americans, and later LGBTQ+ rights, represented the principal promoters of equality.

The constitution of more inclusive egalitarian societies corresponded to a "reaction" yearning to return to a hierarchically regulated era where each human had a defined and crystallized place. An idealized era in which a presumed natural order prevailed, with fixed male and female roles, cultures, and ethnicities considered inferior or superior depending on the hegemonic group's cultural, religious, or technical-scientific viewpoint. To assert racial differences and the hierarchy between roles, cultures, ethnicities, and genders, racists resorted to biological justifications, which led to the racialization of proletarians, women, and homosexuals (Guillaumin, 1995). Racists ultimately shared cultures, discriminating against individuals based on physical or moral qualities. A good example is once again Evola's thought, who considered the Middle Ages a sort of golden age where aristocracy (the second estate) governed; everything after that, from the Enlightenment to modernity, was despised, including emerging social classes.

Racism was therefore also the notion representing a cultural tendency that opposed modern inclusivity values in favor of a past culture that actively promoted discrimination. Racism also massively struck certain cultures and ethnicities, particularly the Jewish one.

Anti-Jewish racism can even be considered the prototype of all other forms of racism. At first, it was religious and spiritual racism against the "deicide race," the killers of Jesus Christ, with anti-Judaism present in Europe for centuries. Spiritual racism was the premise for scientific racism against Jews that spread forcefully in the 1930s and 1940s, involving Germany's ruling class to plan the elimination, one by one, of all Jews worldwide. They were considered a dangerous race, scapegoats of all evil, and bearers of modernity, frightening the bourgeois average man. Jewish intellectuals like Marx, Freud, or Einstein spread ideas revolutionary to past worldviews, scandalizing the conservative majority. Jews were thus targeted by both spiritual and pseudoscientific racists searching for physical signs proving the alleged moral evil projected onto the "deicide race."

For racists, if Jews were "bad," they should also have physiognomies of badness; since medieval times, they were depicted as rapacious and greedy with hooked noses, resembling birds of prey, constructing physical stereotypes, morally placing them on the dark side against the clarity of northern peoples (dark hair vs blond hair, dark eyes vs light eyes).[7]

This represented a systematic dehumanization of Jews by Fascist and Nazi propaganda when racism became a totalitarian state practice, leading to eliminationist policies.[8]

Racisms have subsequently manifested along a kind of geographic spectrum, wherein Nordicism—arguably one of the first organized forms of racism—posited that northern populations were culturally and biologically more evolved and superior compared to southern peoples. Similarly, Eastern populations have consistently been viewed as different and "disturbing" relative to Westerners, a notion encapsulated in Orientalism. At certain historical moments, Eastern populations evoked ambivalent perceptions among Europeans: on one hand, they were seen as bearers of advanced culture and values appealing to Western merchants and elites; on the other, they were regarded as inferior and harmful, much like southern peoples.

Along these geographic axes, various cultures have intersected, clashed, and intermixed. Racisms also represent attempts to halt this intermixture, seen as a kind of disease: a contamination of the dominant culture or biology by dominated populations. Moreover, religious wars, particularistic conflicts, and ethnic wars have been—and continue to be—the most severe manifestations of racist cultures. Thus, racism remains a serious problem for all humanity.

References

Barbujani, G. (2005). Human races: Classifying people vs. understanding diversity. *Current Genomics, 6*, 215–226.

Barbujani, G. (2006). *L'invenzione delle razze: Capire la biodiversità umana* [The invention of races: Understanding human biodiversity]. Bompiani.

[7] The UCEI Antisemitism Observatory website reconstructs how, in Roman times, Jews were identified with the Sabbath (Shabbat, the day of rest) and that Saturday was also the day dedicated to Saturn, a distant and cold planet, a deity with many dark, nocturnal aspects represented by nocturnal birds of prey with hooked beaks. From this arose the myth of the "Jewish nose," which obviously corresponds to no empirical reality and is a prototype of many racist prejudices against Jews (https://www.osservatorioAntisemitismo.it/approfondimenti/e-la-storia-del-naso-adunco/).

[8] Phil Zimbardo (2007) studied such dehumanization techniques as prerequisites to the enactment of evil.

Barbujani, G. (2016). *Gli africani siamo noi: alle origini dell'uomo* [We are the Africans: At the origins of humankind]. Laterza.
Barbujani, G., & Cheli, P. (2022). *Sono Razzista ma sto cercando di smettere* [I'm racist but I'm trying to stop]. Laterza.
Cassata, F. (2003). *A destra del fascismo. Profilo politico di Julius Evola* [To the right of fascism: Political profile of Julius Evola]. Boringhieri.
Curran, A. (2012). Buffon et l'histoire naturelle des Africains [Buffon and the natural history of Africans]. *Dix-huitième siècle, 44*, 183–199.
Guan, M., & So, J. (2023). Social identity theory. In E. Y. Ho, C. L. Bylund, J. C. M. van Weert, I. Basnyat, N. Bol, & M. Dean (Eds.), *The international encyclopedia of health communication* (pp. 1–5). Wiley.
Guillaumin, C. (1995). *Racism, sexism, power and ideology*. Routledge.
Hacking, I. (2005). *Façonner les gens (II). Cours au Collège de France (2004-2005)* [Shaping people (II). Lectures at the Collège de France (2004–2005)]. https://www.college-de-france.fr/fr/agenda/cours/faconner-les-gens-ii
Jackson, J. P., & Weidman, N. M. (2006). *Race, racism, and science: Social impact and interaction*. Rutgers University Press.
La Vergata, A. (2009). *Colpa di Darwin? Razzismo, eugenetica, guerra e altri mali* [Darwin's fault? Racism, eugenics, war, and other evils]. UTET.
Lachman, G. (2018). *Dark star rising: Magick and power in the age of Trump*. TarcherPerigree.
Lauriola, M., Foschi, R., & Marchegiani, L. (2015). Integrating values and cognitive style in a model of right-wing radicalism. *Personality and Individual Differences, 75*, 147–153.
Montuori, A. (2005). How to make enemies and influence people: Anatomy of the anti-pluralist, totalitarian mindset. *Futures, 37*(1), 18–38.
Rota, G. (2003). Un filosofo razzista. Note su Evola [A racist philosopher. Notes on Evola]. *Rivista di Storia della Filosofia, 58*(3), 459–496.
Saini, A. (2019). *Superior: The return of race science*. Beacon Press.
Sidanius, J., Pratto, F., & Devereux, E. (1992). A comparison of symbolic racism theory and social dominance theory as explanations for racial policy attitudes. *The Journal of Social Psychology, 132*(3), 377–395.
Taguieff, P.-A. (1994). *La forza del pregiudizio. Saggio sul razzismo e sull'antirazzismo* [The force of prejudice: An essay on racism and anti-racism]. Il Mulino. (Original work published 1988)
Taguieff, P.-A. (2011, July 15). *L'antiracisme, une machine à exclure?* [Anti-racism, a machine of exclusion?]. Atlantico. https://atlantico.fr/article/decryptage/antiracisme-sos-racisme-concert-pour-egalite-14-juillet

West, C. (1982). *Prophesy deliverance! An Afro-American revolutionary christianity*. The Westminster Press.
Zimbardo, P. G. (2007). *The Lucifer effect*. Random House.

CHAPTER 2

I Am Sorry, But I Am Who I Am, and You Are Not: Spiritual and Cultural Racism

Cultural and religious racism preceded scientific racism. Nevertheless, it is foundational and interwoven with scientific racism.

Racist and discriminatory tendencies of the human mind have ancient psychological roots that project and construct cultures, justifying discrimination based on prejudice. Within cognitive psychology, prejudices primarily represent a blind spot in an individual's ability to perceive and interact with the surrounding world based on conscious, reasoned judgments. Prejudices are also influenced by the specific situations in which an individual is compelled to live. Prejudice is often unconscious and guides behaviors and decisions without being questioned. In this context, prejudice represents a stance toward an individual as a member of a group or an entire set of individuals considered to possess characteristics that are too distant or too different from those of the group to which one belongs: the *ingroup* is automatically perceived as superior, more desirable, and bearer of positive, even moral, attributes; conversely, the *outgroup* is associated with everything the *ingroup* is not. Consequently, there is a tendency to engage in behaviors favoring one's own group to the detriment of others. Minority groups are often the targets of prejudice. These processes can be automatic and, as demonstrated, are often disconnected from any rational justification for differences between groups (Banaji & Greenwald, 2013; Tajfel, 1981; Vincenzo & Foschi, 2021). However, such psychological mechanisms give rise to deeply rooted cultural convictions.

R. Foschi, *Racisms*, https://doi.org/10.1007/978-3-032-22962-5_2

As stated, antisemitism was primarily religious/spiritual racism, upon which further forms of prejudice against the "deicide race" were later superimposed. The Catholic Church officially renounced anti-Jewish prejudice only in 1965 when Pope Paul VI published the declaration of universal brotherhood, *Nostra aetate*, recognizing that every human being could choose a non-Christian religion and concurrently acknowledging the errors committed against Jews, who until then had all been considered responsible for killing Jesus Christ.[1] Paul VI rectified a millennia-old situation that had forced Jews to live in ghettos. In 1555, Pope Paul IV, through *Cum nimis absurdum*, established that Jews bore a guilt that rendered them slaves and promoted anti-Jewish regulations confining and marginalizing them. These measures produced cultural consequences, culminating in centuries of entrenched prejudice that penetrated secular and scientific culture. Jesuits also played a role in this spiritual anti-Jewish racism by associating Jews with the Revolutions of modernity. As recently reconstructed in *La Civiltà Cattolica*, the principal theological journal of the Jesuits:

> To understand the attitude of the ecclesiastical hierarchy and *La Civiltà Cattolica* towards the Jewish question, it is necessary to consider some historical aspects first. From this perspective, a distinction must be made between religious or doctrinal anti-Judaism and socio-political anti-Judaism. The former was due to theological and doctrinal reasons: it viewed the Jew, a man without a homeland, as "damned by God" because of his blindness in not recognizing the Messiah, and his exile was explained through specific religious categories. This included grave accusations of deicide and ritual murder. The dissemination of such ideas was contributed to in the modern era by La Civiltà Cattolica [...]. This anti-Judaic mentality, widespread in broad sectors of European public opinion and not limited to Catholics, condemned Jews to social marginalization. This attitude gave rise to ghettos, instituted in past eras to control Jews under openly discriminatory social legislation (purportedly enacted "more for preventive caution than punitive providence"), but also to protect them against possible widespread pogroms. In any case, although received with Christian charity, the Jew was nonetheless considered an outsider to society.

[1] As highlighted by Di Nola (1973), Pope Paul VI's openings clashed with Western traditionalist culture, which continued promoting versions of spiritual racism already rooted in ideologies disseminated in Europe in the modern era and during the Fascist period.

> Modern anti-Judaism, however, arose with the French Revolution and, in particular, with the social and political emancipation of Jews sanctioned by liberal governments. This liberal legislation […] made Jews "bold and powerful, granting them under the pretext of equality an increasingly predominant status, especially economically, in modern society." Another reason for fighting against the social influence Jews were acquiring, beyond their predominance in economic and financial fields, was the leading role many played in international Freemasonry, which was strongly opposed to Catholicism, and in modern Revolutionary movements, not only in Lenin's Russia but also in Western European nations. This mode of thinking was fueled by the fact that many leaders of European Communist parties were Jews: for example, a majority of the members of the Council of People's Commissars, established by Lenin after the 1917 Russian Revolution as the Revolutionary government of the country, were comprised of Jews. Thus, the figure of the Jew, in Catholic collective imagination and beyond, was assimilated on the one hand to the capitalist exploiting Christian populations, and on the other to the Revolutionary undermining societal foundations. (Sale, 2002, pp. 419–420)

This brief passage masterfully illustrates the birth of the myth of the "demo-pluto-Jewish-Masonic conspiracy," especially prevalent during the Fascist era: the idea that Jews were the cause of liberal democratization, Masonic innovations, and capitalist changes, all regarded negatively in conservative terms. Finally, socialism, though a reaction against capitalism, was also attributed to Jews.[2]

An example of this spiritual racism is found in Agostino Gemelli (1878–1959), an extraordinary promoter of a version of Catholicism open to scientific culture, but who simultaneously rejected sociological and psychological theories such as socialism and psychoanalysis, which he considered to have a common root in Judaism and viewed as a sort of adolescent disease or youthful intoxication. A repentant socialist, Franciscan friar, psychologist, and neo-Thomist, Gemelli established a complex dialogue between science, Catholicism, and the regime during Fascism, often falling into racist arguments, especially when stigmatizing an opponent or a theory that he felt could not coexist with Catholic theology (Foschi et al., 2013, 2018).

[2] "Demo-pluto-Jewish-Masonic conspiracy" was a phrase used by conservatives during Fascism to define an international conspiracy created by wealthy Freemasons, Jews, and communists.

Regarding cultural racism more broadly, Donatella Di Cesare recently analytically highlighted that anti-Jewish racism has been a widely present theme not only in the history of Catholic and Protestant Christianity but also in the history of modern and contemporary philosophy (Di Cesare, 2016). In philosophy, cultural racism is found in many of the Western tradition's "founding fathers." For example, for Di Cesare, racism was a fundamental theme in Georg Wilhelm Friedrich Hegel (1770–1831):

> The political condition of the Jew, who by his very legislation is dispossessed and as a citizen is nothing, has for Hegel also ontological significance. Hence, there is a difficulty in integrating him into the system. Until the end, the Jew remains outside—outside of history, outside of dialectics. Viewed by the Johannine Logos that informs his dialectic and governs the speculative movement, the Jew is—as Derrida says—the "unseen" and "unperceivable," the remnant that does not let itself be permeated, firm and rigid in its resistance. "The Jew is a heart of stone". (p. 59)

Friedrich Wilhelm Nietzsche (1844–1900) represented in the history of philosophy the thinker who first began to consider modernity as decadence, attributing blame to the Jews themselves; similar positions, amplified, are found in the Black Notebooks of Martin Heidegger (1889–1976). Heidegger wrote many statements against "Jews." One of the clearest is the following:

> The Jews 'live' by their marked talent for calculation second only to the principle of race, which is why they are resisting its consistent application with utmost violence. The establishment [Einrichtung] of racial breeding [eugenics] does not stem from 'life' itself, but from the overpowering of life through machination [technology]. What they push forward with such a plan is the complete deracialization of all peoples by clamping them into a uniformly constructed and tailored establishment [Einrichtung] of all beings. At one with deracialization is the self-alienation of peoples – the loss of history – i.e., the decision regions of Be-ing. (Heidegger, 1939–41/2014, 56 [XII, 82–83])

Today, no one would argue that Heidegger was not a supporter of Nazism and that in his ontology, Jews were considered a threat (see Trawny, 2015). Paradoxically, despite the enormous fascist and racist traces left by Heidegger (who signed Sieg Heil in correspondence with explicit Nazi content), until very recently, his exegetes obscured this racism (Ferraris,

2016). Heidegger's case is particularly interesting for understanding the ideological responsibilities of one of the masters of contemporary philosophical thought, who influenced legions of university professors and transmitted a way of thinking that was steeped in racist traditionalism. Although this differed from the scientific and eliminationist racism systematically promoted by Nazism, it nonetheless represented a justification and ideological premise for it. Today, however, even those most interested in Heidegger's thought cannot deny that he was a racist, above all because he rejected the culture represented by the Jews in terms of modernization and a break with tradition (Escudero, 2015; Fuchs, 2015).

As is now well known, prejudice against Judaism in the Black Notebooks fits into a theoretical framework that throughout the twentieth century inspired the "critical" thought of intellectuals both on the right and on the left. Academic philosophy was, however, wrongly portrayed as unrelated to cultural and spiritual racism, which was instead identified only in particular and marginalized figures, such as Julius Evola and Mircea Eliade (1907–1986), both of whom were united precisely by their common rejection of modernity (Piasere, 2021).

Evola's racism can indeed be considered the prototype of cultural and spiritual racism, which has even recently been defined as hyper-racism (Ciracì, 2024). Among his many faults, Evola reinterpreted esotericism in a racist and pro-Fascist manner, reconstructing specific traditional knowledge within conservative and constrictive coordinates that he intended to serve as the cultural foundation of Fascism (see Cassata, 2003; Rota, 2003). Contrary to what Evola often claimed himself, until the early twentieth century, esotericism and Hermeticism, its close relative, had represented a culture of rupture with the conservative and "traditionalist" Catholic tradition [On this point, see numerous writings by Gary Lachman (2008, 2009) on different esoteric approaches relative to progressive versus conservative ideologies]. Embracing Hermetic, Gnostic, and esoteric knowledge across the centuries meant seeking to escape and resist the Inquisition, and thus opening up to Protestantism and modernity. These cultures were prevalent during the Renaissance, the Enlightenment, and the egalitarian Freemasonry of the eighteenth century. They continued to reemerge between the nineteenth and twentieth centuries in many esoteric circles promoting progress, such as Theosophy, an initiatic society that was predominantly female. This progressive, esoteric mindset envisioned a new world built on the values of equality, liberty, and brotherhood. A clear rejection of all distinctions of any kind was fundamental. One principle on which Theosophy was based, for instance, was: "to form a nucleus of the universal brotherhood of humanity, without distinction of race, creed, sex, caste, or colour."[3]

Masonic-esoteric circles were also often international and had a universalistic vision completely opposed to the particularistic and specific vision promoted by Evola in his racist esotericism. Censuring these esoteric-democratic values, Evola sought to turn the same traditions that had promoted innovative visions into racist and conservative ones, replacing modernist esotericism with traditionalist esotericism inspired by Roman paganism but also Oriental culture, which was meant to "brake" the modern world, considered dangerous and degenerate, turning it backward toward the past (Evola, 1934/1995). Evola sought to propose to Fascism a cultural model based on the imperialism of ancient Rome, with the rediscovery of pagan spiritualism, to promote the superiority of Italians over other populations (Evola, 1928/2018). Contrary to internationalist esotericism, the esotericism conveyed by ultraconservatives was thus a system of meanings linked to the creation of national myths and the safeguarding of tradition tied to one's own "lineage." Racism thus inserted itself into an area of interest where knowledge of tradition as a springboard for the future was fundamentally transformed into the promotion of conservation and the rejection of progress.

In his commitment to spreading racism, Evola, for instance, published in 1937 a new translation—initially published in 1921 by the fascist and racist Giovanni Preziosi's (1881–1945) *La Vita Italiana*—of The Protocols of the Elders of Zion. This publication, created from forged documents and distributed in Russia in 1912, attributed responsibility for

[3] The Theosophical Society was founded on three principles; the second and third are as follows: (2) to encourage the study of comparative religion, philosophy, and science; (3) to investigate unexplained laws of Nature and the powers latent in human beings. Helena Blavatsky (1831–1891), founder of Theosophy, promoted a universalist sociability that nevertheless contained a root race theory, which led to controversy over the alleged racism of Theosophists. Among the founders of Theosophy, in fact, there was the idea of a genealogy of all humanity descending from a mythical "Aryan race." Recent criticism of Blavatsky's alleged racism has highlighted that these were complex theories that hybridized biological theories with Eastern cultures, making this "theosophical racism" very different from Western racism (Prophet, 2024). Theosophical Aryanism was Orientalist rather than Nordicist, but in any case, it was later used in German Nazi-Fascist esotericism to justify the racist ideas and practices of that regime, transforming the Aryans from an Iranian to a Nordic people. Theosophists were exploited by German Nazism (see Mosse, 1978), just as esotericism was rewritten in racial terms in Italy, essentially by Evola.

everything representing the modern world to a Jewish-Masonic genealogy that would have utilized the economy and politics to create and govern a new world. The Protocols legitimized the already cited democratic-Jewish-Masonic conspiracy that would have established a modern world opposed to the traditional and even medieval one to which Evola aspired.

Furthermore, all historical phases following the Middle Ages for a racist like Evola represented decadence, the end of the virile and heroic Roman-pagan race that, mixing in subsequent eras, dissolved into feminine (Aphroditic) cultures. The Renaissance society, the Enlightenment, Masonic culture, modernism, and progress were all rejected and considered chaotic (Evola, 1942). Evola's esotericism was thus characterized by a sacred culture of an elitist and hierarchical type that longed for the past and "feared" the future (cf. Foschi, 2024). An aristocratic vision that, although of petty-bourgeois origins, led him to call himself baron to place himself above the multitude. In this sense, his racism was essentially the refusal of all inclusive ideas and therefore also of any religion open to dialogue, as well as the refusal of all classes except the second estate, the aristocracy. Judaism represented precisely the cultural context in which modernity had developed, and for this reason—even while later admitting the falsehood of the Protocols—was condemned by Evola without appeal.

After the catastrophe of World War II, Evola continued to write volumes defending his particular traditionalist and racist conceptions. Above all, he disseminated a pseudo-distinction between his cultural and scientific racism.[4] The latter he considered wrong, an emanation of modernity, and he distanced himself from it. In contrast, the former, also referred to as spiritual, he considered essentially correct but to be distinguished from the scientific one. This self-apologetic attempt was, however, completely meaningless: cultural racism represented and represents the fertile ground on which every other type of racism is built.

It can thus be easily demonstrated that cultural (or spiritual) racism can be considered the context in which scientific racism finds its justification, and that the two forms of racism are substantially interconnected. Moreover, all the ideas on which cultural racism was based—for example, those of Evola—were also present in scientific racism and perhaps formed its

[4] The distinction later became a historiographical category often used to distinguish the two racisms sharply, but it is essentially a misleading distinction.

framework. Above all, the differentialist, hierarchical, and degenerationist conception.

The differentialist premise, according to which a psychological and cultural difference among human beings is taken for granted, is undoubtedly a fundamental principle in both cultural/spiritual and scientific racism; in cultural racism, differences are promoted among elite groups (aristocrats) seen as predestined superiors in hierarchical order who have a sort of dominance over other men. On these grounds, miscegenation and mixing were rejected, with the fear that this would lead to contamination and, ultimately, cultural degeneration, a phenomenon considered typical of democracy and socialism.[5] The fear of degeneration was the obsession of Nazism: in art, for example, classical forms and figures were preferred—in Hitler's paintings, this is evident even to the less expert eye—and informal and abstract art was rejected as degenerate. In 1937, the Nazis even organized a free traveling exhibition of "degenerate art" to propagate among the German people the idea that paintings by prominent avant-garde artists were expressions of madness and depravity.[6]

These principles (differentialist, hierarchical, and degenerationist conceptions) previously present in spiritual racism were actually only reworked and secularized in the scientific context, especially in positivist and late-positivist perspectives. Science attempted to "demonstrate" what cultural racist traditions had already propagated since the Middle Ages.

The case of Eliade's racism is more subtle and ambiguous compared to that of Evola because he himself represented a sort of cultural blind spot model spread during the Cold War. Among intellectuals in the postwar period, support for Fascism and racism by many academics and philosophers of the early twentieth century was easily forgotten, even as they concealed it; it was forgotten that Mircea Eliade had been in the service

[5] The two texts where Evola clarifies his explicitly racist doctrine, trying to distinguish his racism from others, are *Il mito del sangue* [The Myth of the Blood: The Genesis of Racialism] (1937/2017) and *Sintesi di dottrina della razza* [Synthesis of the Doctrine of Race] (1941/2020).

[6] Until the early 1920s, Evola was a Dadaist painter, one of the movements considered degenerate by the Nazis. Evola himself abandoned painting when he developed an interest in philosophy, which then led him to become a traditionalist and racist. It can be hypothesized that the abandonment of Dadaism was internally driven by the conflict between youthful interest in the avant-garde and a subsequent fascination with traditionalist philosophy, which identified the Medieval era as the ideal society—a model certainly not shared by the Dadaists.

of Romanian Fascism and Eliade's later traditionalism was exclusively considered a secular theoretical position as a professor of the history of religions in Paris and Chicago that had nothing to do with his earlier racism and traditionalism.[7] A nostalgic view saw modernity as a catastrophe. In this continuous return to tradition lay the firm conviction that the past held all good, while the indeterminacy of progress was seen as holding all evil; the complex contemporary era was identified as an age of decline, contrasted with an abstract, virtuous, and idealized past. Modernity thus marked a sort of end of history triggered by Jewish culture, which had spread a cultural and spiritual conception from which arose those who built contemporary society and destroyed "tradition." Racist prejudice thus reemerged once again as an idea attributing responsibility for modernity, considered no more and no less than a disaster, to Judaism.[8] Despite many attempts, this Procrustean bed of antisemitic prejudice has never been truly eradicated and reemerges even in recent times in populist movements.[9] Thus, Jews were blamed for all the evils of progress, promoters of freedom of thought, creators of socialism, the most sophisticated sciences, and psychoanalysis:

> An unequivocal sign of a rupture within the fabric of civil society, a dark symptom of historical malaise, of irrationally resolved conflicting tensions, antisemitism accompanies all parallel degenerations of reason: authoritarianism, fascism, Nazism, the squalid monsters born from the "sleep of reason." (Di Nola, 1973, back cover)

True to form, fantastical accusations arose against Jews, such as the "blood libel" spreading from the Middle Ages, whereby Jews were believed to ritually kill and drink the blood of Christian children during

[7] Eliade served Codreanu, the leader of the Romanian fascist party, immediately before World War II. He later concealed this past, which remains important as documented, for example, by correspondence with E.M. Cioran recently published (See Cioran & Eliade, 2019).

[8] The Eliade case resurfaced thanks to Alfonso Maria Di Nola (1926–1997) (See Di Nola, 1977).

[9] Recently, e.g., in Hungary, the right-wing populist government of Orban has repeatedly attempted to close the Hungarian campuses of Central European University, funded by George Soros, a well-known, wealthy Jewish-Hungarian entrepreneur and philanthropist who escaped Nazism and supports progressive (liberal) components of the Democratic Party.

Easter. Despite being paradoxical and utterly false, this accusation was used in different eras to justify racial policies in Europe (Mosse, 1978; Taradel, 2002). From this conspiratorial culture, cultural and spiritual anti-Jewish racism still exists today, constantly redefining itself.

From this viewpoint, it should be remembered that cultural antisemitism, especially in its political-economic form, also involved leftist ideologies. In various ways, they had expressed, since the nineteenth century, anti-Jewish positions in socialist terms. They were supported by such socialist founders as Toussenel and Fourier. This antisemitism was later questioned by socialists themselves, who perceived the danger of racism, which came to be considered one of the degenerations of the conservative European bourgeoisie.

On the other hand, few know that Karl Marx was an admirer of phrenology (Duichin, 2012), the pseudoscience created by Franz J. Gall (1758–1828), widely used by racists to express their conceptions.[10] Indeed, there are numerous passages where Marx expressed positions that support the historical dominance of whites, contrasting with the cultural realities of other ethnicities considered inferior (Moore, 1974). As is well known, in the second part of *On the Jewish Question* (1844), Marx, in response to the critique of Bruno Bauer (1809–1882) against religious attitudes, expressed his own critique of Bauer which contained a broader questioning of Jewish culture with classic anti-Jewish prejudices linked to considering Jews as promoters of modern bourgeois and capitalist society; in this writing, the Jew was seen as an egoist worshiper of money, with the only solution to eliminate religion being, therefore, a Revolution that would also eliminate bourgeois and capitalist society along with its abstract liberal and religious values.

Of course, Marxist ideology was complex and, overall, especially in its twentieth-century critical reading, can be substantially considered as promoting the emancipation of the masses and overcoming societal distortions (and thus also racism). However, as clearly emerges from *The Jewish Question*, before the 1850s Marx regarded Jewish culture as the cause of capitalist modernity, thus necessitating a Revolutionary palingenetic process that would create a society without the contradictions of capitalism and paradoxically emancipate Jews themselves from their culture and religion, considered, in 1844, explicitly disparagingly:

[10] For a description of phrenological theory, see the chapter on scientific racism.

> We are trying to break with the theological formulation of the question. For us, the question of the Jew's capacity for emancipation becomes the question: What particular *social* element has to be overcome in order to abolish Judaism? For the present-day Jew's capacity for emancipation is the relation of Judaism to the emancipation of the modern world. This relation necessarily results from the special position of Judaism in the contemporary enslaved world.
> Let us consider the actual, worldly Jew – not the *Sabbath Jew*, as Bauer does, but the *everyday Jew*.
> Let us not look for the secret of the Jew in his religion, but let us look for the secret of his religion in the real Jew.
> What is the secular basis of Judaism? *Practical* need, *self-interest*. What is the worldly religion of the Jew? *Huckstering*. What is his worldly God? *Money*.
> Very well then! Emancipation from *huckstering* and *money*, consequently from practical, real Judaism, would be the self-emancipation of our time.
> An organization of society which would abolish the preconditions for huckstering, and therefore the possibility of huckstering, would make the Jew impossible. His religious consciousness would be dissipated like a thin haze in the real, vital air of society. On the other hand, if the Jew recognizes that this *practical* nature of his is futile and works to abolish it, he extricates himself from his previous development and works for *human emancipation* as such and turns against the supreme practical expression of human self-estrangement.
> We recognize in Judaism, therefore, a general *anti-social* element of the *present time*, an element which through historical development – to which in this harmful respect the Jews have zealously contributed – has been brought to its present high level, at which it must necessarily begin to disintegrate.
> In the final analysis, the *emancipation of the Jews* is the emancipation of mankind from *Judaism* (Marx, 1844).[11]

Famously, Pierre-Joseph Proudhon (1809–1865), one of the fathers of left-wing anarchism, already expressed anti-Jewish views in some newspapers and wrote a note in his diaries in 1847 explicitly considering Jews (including Marx) as founding enemies of all economic perversions linked to capitalism:

[11] https://www.marxists.org/archive/marx/works/1844/jewish-question/.

> December 26, 1847: Jews. Write an article against this race that poisons everything by sticking its nose into everything without ever mixing with any other people. Demand its expulsion from France with the exception of those individuals married to French women. Abolish synagogues and not admit them to any employment. Demand its expulsion Finally, pursue the abolition of this religion. It's not without cause that the Christians called them deicides. The Jew is the enemy of humankind. They must be sent back to Asia or be exterminated. H. Heine, A. Weill, and others are nothing but secret spies; Rothschild, Crémieux, Marx, Fould, wicked, bilious, envious, bitter, etc., beings who hate us. The Jew must disappear by steel or by fusion or by expulsion. Tolerate the elderly who no longer have children. Work to be done – What the peoples of the Middle Ages hated instinctively I hate upon reflection and irrevocably. The hatred of the Jew like the hatred of the English should be our first article of political faith. Moreover, the abolition of Judaism will come with the abolition of other religions. Begin by not allocating funds to the clergy and leaving this to religious offerings. – And then, a short while later, abolish the religion. (Proudhon, 1847, vol. 2, p. 337)[12]

Such racism was shared in the nineteenth century by many other leftist intellectuals who saw Judaism as the cause of the capitalist crisis produced by modernity, and whose extermination plan, as prefigured by Proudhon, a libertarian leftist ideologue, in this cruel diary page, would paradoxically become the backbone of right-wing antisemitism during the nazi era. Although, as recently demonstrated (Dreyfus, 2011), explicit left-wing antisemitism progressively limited itself to some radical groups, in the nineteenth century, it was widespread also within the labor movement despite (self-)critical positions already expressed by Friedrich Engels (1820–1895) (see Engels, 1890/2002).

A lengthy debate arose on left-wing racism and antisemitism, mainly because Marx and Engels, from the 1850s onward—probably in reaction to the contradictions of racist anthropology, as well as the impossible defense of racist positions within a humanitarian socialist vision—renounced antisemitism. Moreover, after Marx's death, the labor movement redefined the issue, considering antisemitism a dangerous bourgeois prejudice elaborated in response to its crisis.

[12] https://www.marxists.org/reference/subject/economics/proudhon/1847/Jews.htm.

The infamous Dreyfus affair marked a pivotal moment that reshaped the Jewish question on the left, prompting its reevaluation (Burgio & Gabrielli, 2012; Dreyfus, 2011). At the end of the nineteenth century, French army captain Alfred Dreyfus (1859–1935) was accused of being a spy for the Germans. In 1894, based on forged evidence, the government and the army chief of staff identified the Jewish Dreyfus as the ideal culprit of an information leak to the German army. Thus emerged a deeply racist French society unwilling to believe in Dreyfus, who, due to his Jewish origin, could even be sacrificed for state reasons (Fig. 2.1). Dreyfus began a harsh imprisonment on a small island in French Guiana (Devil's Island).

Despite the true culprit of the leak (officer F. W. Esterhazy) being found, the trial lasted twelve years, and Dreyfus was fully exonerated only at the end of this long period. France, in the whole Third Republic—a legendary period of secular state reorganization following the Paris Commune (1871) and lasting until the Nazi invasion of 1940—was divided into "Dreyfusards" and "anti-Dreyfusards." At the beginning of this saga, the majority opposed Dreyfus, many believing the army's investigation; even the proletariat was mainly against the Jew, considered a scapegoat for its misfortunes and a class enemy. The anti-Dreyfusards were mainly composed of people in power (nationalists, conservatives, and the military). The case began to change thanks to Colonel M. G. Picquart, who was elected head of a secret services section and intercepted documents proving Dreyfus's innocence. Despite facing hostility from his superiors, Picquart started a secular campaign within the army to rehabilitate Dreyfus and condemn the real spy. Over the years, public opinion became increasingly involved in the case, and influential intellectuals, including socialists and republicans such as Émile Zola (1840–1902), organized a campaign in favor of Dreyfus. In 1898, Zola wrote the article "J'Accuse…!" in the newspaper *L'Aurore*, accusing the army leadership of deliberately condemning Dreyfus based on fabricated evidence and gross errors that authoritatively persevered even in the face of opposite evidence. After the Dreyfus affair, the Third Republic shifted toward radical-socialist governments; France and the labor movement undertook a virtuous process in which anti-Jewish prejudice was increasingly condemned and marginalized. Progressives prevailed, and Dreyfus was pardoned by Émile Loubet (1838–1929), President of the Republic and a radical republican elected in 1899; subsequently, Dreyfus was also judicially rehabilitated. The Dreyfus case revealed a Europe in which antisemitism was nonetheless rampant and organized.

Fig. 2.1 Dehumanizing caricature of the "traitor" Dreyfus depicted as a snake

From the 1890s, also following the case's resolution and the emergence of a Jewish working class, the European left became increasingly aware of the danger of anti-Jewish prejudice that could nest even in anti-capitalist tradition. The problem of left-wing antisemitism remains a historiographic open question engaging many scholars with non-univocal interpretations.[13] The dual attitude of the left toward Jews continued; in the twentieth century, on one hand, figures like Di Nola, a fundamental figure in Italian anthropology and a Marxist intellectual, explicitly highlighted the perniciousness of longstanding and varied anti-Jewish racism, both Catholic and leftist. In contrast, others sought to downplay this "democratic" antisemitism, which was nonetheless characterized in various ways and survived into contemporary times mainly as anti-Zionism (Battini, 2010; Schirripa, 2021; Tarquini, 2019).[14]

Since the times of Dreyfus, we have been facing cultural racism built on a conspiracy mentality; cultural and spiritual anti-Jewish racism thus continues to appear and reappear today. For example, political opposition to Israeli governments is automatically generalized to all Jews, making them responsible for the actions of a state for which they are not objectively responsible and with which they identify according to different standpoints (Meghnagi, 2004); just as once they were all considered Jesus Christ's killers, today Jews are often considered, without distinction, supporters of Israeli policy toward Palestinians.

Spiritual and cultural racism over the centuries has also been widespread and present, not only directed against Jews. Every empire or hegemonic society of the past developed cultural racist theories, morally justifying domination over subjugated peoples. The Spanish Empire is one of the most systematic examples where racism produced the destruction of entire South American cultures based mainly on religious or moral justifications. If the conquered peoples were not Catholic and retained their moral and spiritual conceptions, they were easily considered primitive and worthy of assimilation and enslavement by those who felt superior

[13] The example most debated by historians was anti-Semitism in the Soviet Union. Although many Bolsheviks were of Jewish origin and anti-racist Soviet laws were enacted after the Russian Revolution of 1917, over time, there was a growing ambiguity toward Jews due to their cosmopolitanism and refusal to adapt to collectivization and Sovietization, especially during the Stalinist era. Stalin first established a territory in the Far East for Jews (Jewish Autonomous Oblast) and then persecuted them more and more openly (Vaksberg, 1994).

[14] Zionism is a political movement born at the end of the nineteenth century whose mission was the creation and consolidation of the State of Israel, a territory to accommodate persecuted Jews worldwide in their lands of origin. This has, as is well known, generated a protracted conflict with Palestinian Arabs occupying the same territories (Edelheit, 2019).

and chosen by God for a forced civilizing mission that often resulted in genocide.

Similarly, European colonizers in what would become the United States first displaced and decimated Indigenous American populations, enslaved Africans, and subsequently employed eugenic ideologies to control and categorize African Americans and impoverished European immigrants. Cultural racism was also deeply embedded within the imperialist ideologies of the English, French, and Russian empires, which consistently maintained varying levels of discriminatory and often aggressive attitudes toward colonized peoples. Likewise, Italians exhibited racist practices toward Abyssinians and Libyans. In sum, this represents a long and repetitive historical pattern from which humanity has found it exceedingly difficult to emancipate itself. Even the most recent xenophobia against Islamic immigrants present in all Western countries, defined as Islamophobia, reactivates old religious conflicts that date back to the medieval crusades.

In general terms, Western colonialism considered conquered cultures to be different, bizarre, inferior, but also, in some cases, mystical or bearers of unknown and magical knowledge. In any case, the relationship between the West and the East has been characterized by prejudice that has been defined as Orientalism (Said, 1978).

This pervasive racism, however, was not a phenomenon exclusive to the West. Across global history, numerous non-European societies similarly asserted their own cultural superiority over neighboring groups, thereby justifying policies of exploitation and even genocide. A less widely recognized instance is the racism embedded within Japanese culture toward their neighboring Chinese, Koreans, and other proximate populations, whom Japan historically governed or engaged in conflict against. The traditional spirit and values of Japan—referred to as Yamato-damashii—were regarded as inherently superior and constituted a legitimate rationale for exercising authoritarian control over other Asian peoples.

On the cultural and religious front, India represents another distinct case of persistent racism. The country continues to uphold a hierarchical caste system structured around four primary castes, with an additional, separate fifth caste known as paria or Dalit, often referred to as "untouchables" or an oppressed group. The first four Indian castes (priests, warriors, merchants, and servants) are traditionally considered emanations of different parts of the primordial deity Puruṣa's body. In contrast, the Dalits are considered outside the caste system, regarded as lower than servants, and not born as emanations of the deity. This system produces a rigid aristocratic social stratification where individuals assigned to lower castes or classes are deprived of the freedoms and resources granted to those in higher castes or classes. Laws enacted to abolish the caste system continue to face significant challenges in enforcement due to deeply ingrained spiritual and traditional religious beliefs that support and legitimize the division and hierarchical structure of Indian society.

References

Banaji, M. R., & Greenwald, A. G. (2013). *Blindspot: Hidden biases of good people*. Delacorte Press.

Battini, M. (2010). *Il socialismo degli imbecilli. Propaganda, falsificazione, persecuzione degli ebrei* [The socialism of fools: Propaganda, falsification, and persecution of Jews]. Bollati Boringhieri.

Burgio, A., & Gabrielli, G. (2012). *Il razzismo* [Racism]. Ediesse.

Cassata, F. (2003). *A destra del fascismo. Profilo politico di Julius Evola* [To the right of fascism: Political profile of Julius Evola]. Boringhieri.

Cioran, E. M., Eliade M. (2019). *Una segreta complicità*. Lettere 1933–1983 [A Secret Complicity. Letters 1933–1983]. Adelphi.

Ciracì, F. (2024). *Sul razzismo. Strutture logiche e paradigmi storico-filosofici* [On racism: Logical structures and historical-philosophical paradigms]. Mimesis.

Di Cesare, D. (2016). *Heidegger e gli ebrei: I «Quaderni neri»* [Heidegger and the Jews: The "Black notebooks"]. Bollati Boringhieri.

Di Nola, A. M. (1973). *Antisemitismo in Italia 1962/1972* [Antisemitism in Italy 1962/1972]. Vallecchi.

Di Nola, A. M. (1977). Mircea Eliade e l'antisemitismo [Mircea Eliade and antisemitism]. *La Rassegna Mensile di Israel, 43*(1/2), 12–15.

Dreyfus, M. (2011). *L'antisémitisme à gauche. Histoire d'un paradoxe de 1830 à nos jours* [Anti-semitism on the left: The history of a paradox from 1830 to the present day]. La Découverte.

Duichin, M. (2012). Marx «frenologo». Una pagina rimossa di storia delle idee [Marx "phrenologist": A removed page from history of ideas]. *Intersezioni, 2*, 211–234.

Edelheit, H. (2019). *History of Zionism: A handbook and dictionary*. Routledge

Engels, F. (2002). *Über den Antisemitismus* [About anti-Semitism]. In *Werke, Artikel, Entwürfe, Oktober 1886 bis Februar 1891* [Works, articles, drafts, october 1886 to February 1891] (vol. 31, pp. 1033–1037). Akademie Verlag (Original work published 1890).

Escudero, J. A. (2015). Heidegger's Black notebooks and the question of anti-semitism. *Gatherings: The Heidegger Circle Annual, 5*, 21–49.

Evola, J. (1942). Il miraggio del Rinascimento [The mirage of the renaissance]. *Carattere, Rassegna Del Lavoro Italiano, 2*, 5–6.
Evola, J. (1995). *Revolt Against the Modern World*. Inner Traditions International. (Original work published 1934).
Evola, J. (2017). *The myth of the blood: The genesis of racialism*. Arktos (Original work published 1937).
Evola, J. (2018). *Pagan imperialism and metaphysics of war*. Wewelsburg archives publication (Original work published 1928).
Evola, J. (2020). *Synthesis of the Doctrine of Race*. Cariou Publishing (Original work published 1941).
Ferraris, M. (2016). Il complotto di Heidegger [The Heidegger conspiracy]. *Rivista di Estetica, 62*, 133–142.
Foschi, R. (2024). "He who leaves the old way for the new, knows what he leaves, but does not know what he will find": Traditionalism, zero-sum thinking, and imagined future. *World Futures, 80*(3), 223–240.
Foschi, R., Giannone, A., & Giuliani, A. (2013). Italian psychology under protection: Agostino Gemelli between Catholicism and fascism. *History of Psychology, 16*(2), 130–144.
Foschi, R., Innamorati, M., & Taradel, R. (2018). «A disease of our time»: The Catholic Church's condemnation and absolution of psychoanalysis (1924–1975). *Journal of the History of the Behavioral Sciences, 54*(2), 85–100.
Fuchs, C. (2015). Martin Heidegger's anti-semitism: Philosophy of technology and the media in the light of the "Black notebooks". Implications for the reception of Heidegger in media and communication studies. *TripleC: Communication, Capitalism & Critique. Open Access Journal for a Global Sustainable Information Society, 13*(1), 55–78.
Heidegger, M. (2014). *Gesamtausgabe, Band 96: Überlegungen XII-XV* [Complete Edition, Volume 96: Reflections XII–XV]. Klostermann (Original work published 1939–1941)
Lachman, G. (2008). *Politics and the occult: The left, the right, and the radically unseen*. Quest Books.
Lachman, G. (2009). *A dark muse: A history of the occult*. Basic Books.
Marx, K. (1844). *On the Jewish question*. https://www.marxists.org/archive/marx/works/1844/jewish-question/
Meghnagi, D. (2004). A proposito del convegno dell'UCEI tra Vecchio e nuovo antisemitismo [Regarding the UCEI conference on old and new antisemitism]. *La Rassegna Mensile di Israel, 70*(3), 185–189.
Moore, C. (1974). Were Marx and Engels white racists? The prolet-aryan outlook of Marxism. *Berkeley Journal of Sociology, 19*, 125–156.
Mosse, G. L. (1978). *Toward the final solution: A history of European Racism*. Fertig.

Piasere, L. (2021). *Antisemitismo e antiziganismo in Italia. Sull'antropologia del razzismo di Alfonso Maria di Nola* [Antisemitism and antiziganism in Italy: On the anthropology of racism of Alfonso Maria di Nola]. Morlacchi.

Prophet, E. (2024). *Racist or liberating? The difficult transformations of theosophical root race theory.* In C. M. Stang & J. A. Josephson Storm (Eds.), *Theosophy and the study of religion* (pp. 145–173). Brill.

Proudhon, P.-J. (1847). *On the Jews.* https://www.marxists.org/reference/subject/economics/proudhon/1847/jews.htm

Rota, G. (2003). Un filosofo razzista. Note su Evola [A racist philosopher: Notes on Evola]. *Rivista di Storia della Filosofia*, *58*(3), 459–496.

Said, E. (1978). *Orientalism.* Pantheon Books.

Sale, G. (2002). Antigiudaismo o antisemitismo? Le accuse contro la chiesa e la «Civiltà Cattolica» [Anti-Judaism or antisemitism? Accusations against the church and Civiltà Cattolica]. *Civiltà Cattolica, Quaderno, 3647*(II), 419–431.

Schirripa, P. (2021). L'urgenza di un'antropologia dell'antisemitismo [The urgency of an anthropology of antisemitism]. In L. Piasere (Ed.), *Antisemitismo e antiziganismo in Italia. Sull'antropologia del razzismo di Alfonso Maria di Nola* [Antisemitism and antiziganism in Italy. On the anthropology of Alfonso Maria di Nola's racism] (pp. 107–113). Morlacchi.

Tajfel, H. (1981). *Human groups & social categories.* Cambridge University Press.

Taradel, R. (2002). *L'accusa del sangue: storia politica di un mito antisemita* [The blood accusation: Political history of an anti-Semitic myth]. Edizioni Riunite.

Tarquini, A. (2019). *La sinistra italiana e gli ebrei. Socialismo, sionismo e antisemitismo dal 1892 al 1992* [The Italian left and the Jews: Socialism, Zionism, and anti-Semitism from 1892 to 1992]. Il Mulino.

Trawny, P. (2015). *Heidegger and the myth of a Jewish world conspiracy.* University of Chicago Press.

Vaksberg, A. (1994). *Stalin against the Jews.* Knopf.

Vincenzo, C., & Foschi, R. (2021). *Razzismo vecchio stile, razzismo moderno, populismo: una ricerca di psicologia sociale* [Old style racism, modern racism, populism: A social psychology study]. In R. Genovese (Ed.), *Sulla sindrome identitaria* [On the identity syndrome] (pp. 23–70). Rosenberg & Sellier.

CHAPTER 3

On Scientific Racism

Spiritual racism represented a pervasive culture that sought to dominate other cultures from an anti-democratic, elitist, aristocratic, and imperialist viewpoint. Some, like Evola, attempted to separate a spiritual/cultural conception of racism from a scientific one and even justified their stance by considering only the theories of scientific racist scientists as truly racist. As we saw in the previous chapter, even the anti-Jewish positions of important philosophers such as Heidegger are downplayed by his exegetes because they are considered different from biological racism.

In reality, between the eighteenth and twentieth centuries in the West, cultural and spiritual racism coexisted and hybridized with various pseudoscientific theories that sought to justify racism by systematizing differences among people and groups through various measurements. Many scientists and intellectuals were steeped in spiritual racism and viewed racial hierarchies as self-evident truths. They saw the world through prejudiced lenses.

The intertwining of cultural and scientific racism makes it very difficult to judge posthumously how correct it is to define as racist figures who lived in an era when it was taken for granted that human groups were classified into superior and inferior races. However, among scholars, some positions placed racist, fixist, and essentialist conceptions at the center of their theories. In contrast, others viewed discrimination as marginal or mitigated by Lamarckian evolutionary approaches, in which people and generations were considered not fixed but modifiable over time through

R. Foschi, *Racisms*, https://doi.org/10.1007/978-3-032-22962-5_3

experience and education. In any case, a significant number of scientists openly embraced racist views. Cataloging the scientists who exhibited overt or subtle discriminatory positions from the eighteenth century onward is an onerous task.

Scientific doctrines also contained sexist theories that regarded males as biologically superior. Racist and sexist scientists wrote similar things: they attempted to provide a "pseudoscientific" demonstration that human races existed in hierarchical order, that psychological differences existed among them, and that people had different skulls and bodily features that justified a hierarchical vision of humanity. Northern populations were generally viewed as morally superior to southern populations, who were seen as more mixed and psychologically inferior. Additionally, children were considered inferior to adults, and women were considered inferior to men. Some even drew lines of division in east-west terms. As described in the previous chapter, eastern populations similarly held distinct and often negative biases (Said, 1978).

Even early Enlightenment thinkers applied the concept of equality mostly within their own "race," assuming natural hierarchies among human races as a self-evident fact. Founders of the Enlightenment and modern thought, such as Voltaire (1694–1778), did not escape this paradigm. Notably, racial classifications and hierarchies were even pursued for aesthetic and artistic purposes, as exemplified by anatomist Petrus Camper (1722–1789). He is known for having imagined that the facial angle produced by two straight lines, one passing through the forehead and the other through the ears, was of some use in racial classification (Fig. 3.1). Their intersection at the nose formed an angle. The more it was acute, the more it indicated facial perfection and beauty. This in turn corresponded to greater intelligence and moral beauty. According to Camper, black people had an obtuse angle. Based on the facial angle, actual differential hierarchies were established according to the facial inclinations of different individuals; the facial angle became a precursor to intelligence tests, considered valuable as a measure of psychological and moral identity (Gould, 1987, 1996).

Scientific racism then became intertwined with slavery. If a race existed that presented biological characteristics that were undeveloped and animalistic, it would be possible to justify legislation establishing slavery. Just as animals were used for work, races considered closer to the animal world could be exploited for similar purposes. Among white people, a belief arose, based on pseudoscientific racial concepts,

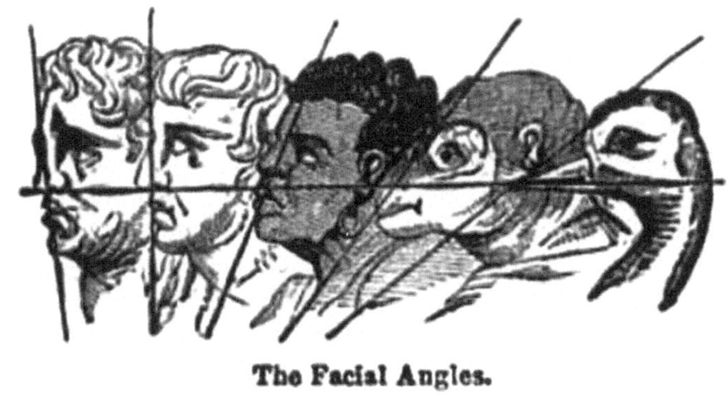

Fig. 3.1 Facial angles. From The National Encyclopaedia, vol. 5, ed. W. Mackenzie, Oxford, 1875, p. 406

that slavery was lawful because slaves were not considered members of "humanity" with the same rights and privileges as white people. However, a rift soon developed with those who did not believe in this slave-owning concept.

In nineteenth-century Europe, the biological study of individual differences was also widespread; these differential studies heralded further articulations of biological racism, particularly during the late nineteenth century, a period marked by the rise of positivism. At the beginning of the nineteenth century, Franz Joseph Gall's phrenology appeared to be part of modern science. For Gall, the brain was divided into organs that controlled people's morals and psychology. By studying the shape of the skull, it was possible to gain insight into individual differences, as the skull served as a sort of map of the size and shape of the brain organs it contained. Phrenology promoted the idea that human groups could be classified according to obvious craniological characteristics.

For his part, Johann Friedrich Blumenbach (1752–1840), one of the most systematic German anthropologists of the early nineteenth century, despite being against slavery and not being a polygenist, provided a classification for these supposed human races. His last ranking was composed by five groups based on skin color correlated with other physical and craniological characteristics: white or Caucasian, black or Ethiopian (African), yellow or Mongolian, olive or Malay, red or American (Blumenbach, 1775, 1795/1865). Blumenbach did not believe that races were fixed entities but that there was a certain variability between them. However, the classification itself lent itself to simplification and became a prototype for all subsequent pseudoscientific racists (Gould, 1994). Physiognomy and phrenology were thus among the primary tools used to demonstrate the supposed reality of races and justify racism on (pseudo)scientific grounds. Naturalistic conceptions about crania differing in blacks and the so-called use of the facial angle aimed to identify inferior or mentally weak individuals. Such ideas were taken for granted and formed the basis for anthropological racist thought found in American naturalists who applied European classifications for explicitly racist purposes.

Among the first scientists to adopt an explicit theory of race were Robert Knox (1793–1862), James Hunt (1833–1869), Samuel George Morton (1799–1851), and Louis Agassiz (1807–1873), most of whom were polygenists and creationists. Their common belief, based on craniological measurements and morphological classifications, was that human races existed, and that they were distinct and hierarchical in nature. Whites were deemed morally superior to others—red, yellow, or black—blending scientific and religious justifications (La Vergata, 2009). Knox wrote, "Race is everything: literature, science, art—in a word, civilization, depends on it" (Knox, 1862, p. V). Imagining immutable races, studying their bones and environments, polygenist racists sought to identify essential characteristics that hierarchically marked superior races.

Then the books of the most radical racists appeared in Europe: Joseph Arthur de Gobineau (1816–1882), Georges Vacher de Lapouge (1854–1936), and Houston Stewart Chamberlain (1855–1927).

All three are sadly notorious for having written treatises that were not particularly original and gave rise to clusters of ideas and ideologies that became the basis for racism and prejudice in all its forms (Nordicism, antisemitism, etc.). These "fathers of racism" firmly believed that skin color objectively represented the existence of races (white, black, yellow, red, olive), that mixing would lead to degeneration, and that white people were somehow descended from a mythical superior and pure Iranian lineage: the "Aryans" inhabiting the Middle East and India. It was mistakenly believed that the "Aryans" were the progenitors of the "aristocrats" in analogy with the root "ari,"[1] and that their

[1] "Aryan" actually referred to Indian languages, not aristocrats. Cf. Cavalli-Sforza and Cavalli-Sforza (2013, p. 360).

physical qualities were linked to their moral qualities and together constituted the essence of this mythical superior race (Chamberlain, 1899; de Gobineau, 1853–1855; de Lapouge, 1899). De Gobineau was considered the first systematic racist and the one who influenced Hitler in his delusions about Aryanism, Nordicism (for these racists, the Aryans were mostly identified with the Anglo-Saxons, who were descended in relatively obscure ways from this Iranian lineage),[2] and racial purity. That said, I do not believe it is useful here to trace a history that dwells on summarizing the thinking of these systematic racists, perhaps highlighting the few differences between them. It is worth noting that Nordicism was central to de Gobineau, but not the anti-Jewish racism found in de Lapouge and Chamberlain. These advocates of discrimination and racism were clearly contradictory and represented only the tip of the iceberg of racists. De Lapouge went so far as to write that the motto of the French Revolution, "Liberty, Equality, and Fraternity," should be replaced with "determinism, inequality, and selection" (Taguieff, 2005). For them, racism was therefore also a counter-revolutionary ideology. For these explicitly racist scholars, the boundary between cultural and scientific racism is nonexistent, and the two forms of racism reinforce each other.

Having introduced the problem of scientific racism, it is interesting to highlight the historiographical categories in which this racism can be understood. As already pointed out, scientists have traditionally sought racial types on which, however, they disagreed. Polygenists were mainly the first racists, but there were also racists among creationists and monogenists. Even Charles Darwin (1809–1882), who based his theory of evolution on monogenism, was convinced of the racial differences between civilized peoples and the savages he encountered on his travels. In *The Descent of Man*, he described in various passages the races of uncivilized peoples he had observed. Darwin was convinced, like other naturalists of his generation, that despite efforts, it was impossible to civilize certain races (in particular, historians often cite the episode described by Darwin concerning some inhabitants of Tierra del Fuego who were

[2] The origin and development of the Aryan myth is quite interesting because the identification of Aryans with Nordics was unproven yet dogmatically assumed. Among others, this identification of Aryans as a people of sages existed even in cultures like Theosophy, which opposed racism and aspired to universal brotherhood.

brought to England and educated to become missionaries in their homelands. When they were returned to their homeland, they dispersed and returned to living according to the customs of their original community) (Darwin, 1871).

Darwin and the early Darwinists, however, by introducing a dynamic system of natural selection[3] of species that also took environmental pressures into account, overcame fixism and polygenism,[4] but gave rise to a Darwinian racism that attributed the development of different races not to different ancestors, but to evolutionary processes. In this sense, due to natural selection and environmental pressures, evolution would have produced human groups so different that they would eventually be associated with actual races. Polygenism, therefore, was not entirely overcome but remained at least conceptually hidden in a series of anthropological theories based on measurements of the skull and body, which nevertheless led to differential typologies that attempted to describe human hierarchies with types and subtypes of northern populations and types and subtypes of southern populations.

In the second half of the nineteenth century, countless anthropologists followed these lines of research (Jackson & Weidman, 2006, pp. 71–76). Evolutionism, which referred to Darwin's natural selection, then influenced the development of social and psychological theories. Herbert Spencer (1820–1903) developed the concept of the survival of the fittest, individuals who possessed the most suitable characteristics. Spencer and those who followed a similar line of thought, such as Ernst H. Haeckel (1834–1919), believed that individuals and societies were greatly influenced by evolution and could be in a more or less civilized state. A mind could therefore be considered on a spectrum from primitive to evolved. For Spencer, individuals and races were placed on an evolutionary scale. Not only facial angle, but also more coarse measurements such as brain mass, indicated greater or lesser evolution of human groups, as well as differences between children and women.

[3] Alfred Russel Wallace (1823–1913) is known for co-discovering natural selection with Darwin, though with a somewhat different general outlook. He remained on good terms with Darwin.

[4] The main opponents of Darwinian evolution were creationists and polygenist racists such as Agassiz. Conversely, polygenists like Morton and Knox were criticized by Darwin himself.

The Swedish anthropologist Anders Retzius (1796–1860) developed a method of measuring skulls that became widely adopted among anthropologists. He contrasted people with narrow heads (brachycephalics) with those with long heads (dolichocephalics). Retzius and many other anthropologists incorporated these head measurements into their racial theory, and "those with long heads" became the prototype of the superior Nordics as opposed to others. However, the measurements involved many ambiguous and difficult-to-classify situations, so much so that Paul P. Broca (1824–1880), one of the fathers of French anthropology, devised less crude measures such as the cranial index, which was the ratio between the width and length of the skull, but failed to ascertain the superiority of dolichocephalics over brachycephalics. Others, such as De Lapouge, identified Aryans with blond dolichocephalics, while southern populations had darker skin, were inferior, and brachycephalic. De Lapouge created the myth of using skull length as an objective measure for racist purposes. All the most important figures, despite all evidence to the contrary, were thus considered Aryans, including Jesus Christ.

These were ways of capturing the differences between the bodies of various ethnic groups and racializing populations. One problem was that these measurements never corresponded to the researchers' prejudices, so that, for example, the French, apart from Lapouge, who was consistently pro-Nordic, did not accept that the Germans were more dolichocephalic (Gould, 1996).

These ideas involved society in general, which was influenced by the rules of evolution, such as the survival of the fittest; positivist scientists sought signs of greater or lesser evolution in the subjects they studied. From this perspective, Broca's case marked the failure of anthropology to identify genuine hierarchical differences. In fact, as demonstrated by Cryle (2025), while believing that human groups—divided on the basis of skin color—were more or less morally "perfectible," he admitted the impossibility of distinctions between dolichocephalics and brachycephalics based on their presumed greater or lesser capacity for civilization.

The concept of the primitive or the evolutionary hierarchy of subjects also lent itself to new forms of discrimination that could affect categories other than "savages" (Jackson & Weidman, 2006, pp. 80–83). Children, viewed as being at an early stage of development, were considered less evolved than adults, women were believed to be less evolved than men, and the "insane" were considered primitive like women, children, and savages. Criminals and crowds, considered to be associated minds, were also believed to have regressed to a "primitive" stage of mental development. Racist concepts in this evolutionism thus became a sort of application of an idea of progress/regression, evolved/primitive, which affected various types of people: women, the mentally ill, children, "savages," criminals, and finally the fourth estate (the proletariat), considered a dangerous mob and therefore primitive by citizens belonging to the first estate (the clergy), second estate (the aristocracy), and third estate (the bourgeoisie).

Among the Italian scientists who, as we shall see, can be classified within this concept strongly characterized by an evolutionism that hierarchically classified individuals and human groups, specific reference must be made to Cesare Lombroso (1835–1909).

Lombroso was primarily concerned with crime, but he did not shy away from racist and hierarchical attitudes toward women and different ethnic groups. For Lombroso, criminals, women, and savages were characterized by "atavism," a concept he introduced to justify the primitive state of some individuals or groups compared to others.

Thus, crowds were also represented as regressed, emotional, and savage, and a conservative and classist mentality characterized crowd psychologists.[5] Specifically, some pioneers in the study of crowds, at the time when psychology was becoming independent from philosophy and physiology, and based on this "defensive" evolutionism, favored the spread of conservative ideas within the new positive human sciences.

The Paris Commune of 1871 marked an important transition because positivists saw it as the realization of the danger that power could fall into the hands of a savage and regressed mob. The French bourgeoisie, which included the first psychologists and criminologists, was concerned about the behavior of the mob, and its representatives began to address the issue. Even Hippolyte A. Taine (1828–1893), a positivist historian who was among the first to conceive epistemologically the need for scientific psychology, emphasized the danger of the crowd, which was emerging as a new subject in history (Van Ginneken, 1992). Within the Italian positive school of criminal law, born from the ideas of Lombroso and his followers, Scipio Sighele (1868–1913) developed his own psychology of crowds, drawing a parallel between criminals and "delinquent crowds" to justify the crimes that took place during mass gatherings or collective

[5] To my knowledge, the only crowd psychologist who viewed the mentality of crowds positively was the Calabrian Pasquale Rossi (1867–1905).

protests, several years before Gustav Le Bon (1841–1931), the most cited of the early crowd psychologists. In his 1891 book *La folla delinquente* (The Delinquent Crowd), written at a time when the labor movement was organizing and proletarian demonstrations were viewed with fear by the bourgeoisie, Sighele believed that the behavior of the masses was conditioned by hereditary criminal factors that emerged simply because large groups of people were gathered together. Sighele thus expressed a negative view of crowds, tending to criminalize associative and mass phenomena. Sighele's political trajectory, from socialism to nationalism and then to fascism, also demonstrated the elitism inherent in collective psychology, which, fundamentally, with proletarian and sub-proletarian demonstrations in mind, considered the working class to be dangerous, regressive, and in need of control and subjugation. Workers were considered a primitive population, belonging to a different and hierarchically inferior social class, akin to a distinct race.

Among all crowd psychologists, Le Bon was perhaps the one who best represented the prototype of the conservative positivist. He was primarily a medical anthropologist and student of Broca. He undertook an intensive and differential study of ethnic groups, using the tools of craniometry as a doctor in the French colonies. In his homeland, he became a pioneer of crowd psychology and political psychology. He was an author read and appreciated by Hitler and Mussolini. The Frenchman was also influential in Freud's psychology of the masses and in the theory of propaganda as an effective tool for controlling the population in both democratic and authoritarian regimes (Van Ginneken, 1992).[6]

He firmly believed that crowds had a regressed mentality, like savages and women, and that, like both, crowds should be tamed by the elite through political and communication strategies or by using regressive means such as racism. Racism and sexism were the hallmarks of Le Bon's beliefs. Regarding women's emancipationist demands, he expressed himself as follows:

> The desire to provide them with the same education and, consequently, to set them the same goals [as men] is a dangerous chimera [...] The day they fail to understand that nature has destined them for inferior occupations, women will leave their homes and join our battles; that day will mark the

[6] Edward Bernays (1891–1995), Freud's nephew, was widely credited as the modern inventor of propaganda as a way to manage and control the masses (Bernays, 1928).

beginning of a social revolution, and everything that preserves the sacred bonds of the family will disappear. (Le Bon, 1879, p. 62)

References

Bernays, E. L. (1928). *Propaganda*. Liveright.

Blumenbach, J. F. (1775). *De generis humani varietate nativa* [On the natural variety of the human species]. apud Vandenhoek et Ruprecht.

Blumenbach, J. F. (1865). *The anthropological treatises of Johann Friedrich Blumenbach* (T. Bendyshe, Trans. and Ed.). Anthropological Society. (Original work published 1795).

Cavalli-Sforza, L. L., & Cavalli-Sforza, F. (2013). *Chi siamo: la storia della diversità umana* [Who we are: The history of human diversity]. Codice edizioni.

Chamberlain, H. S. (1899). *Die Grundlagen des Neunzehnten Jahrhunderts* [The foundations of the nineteenth century]. Bruckmann.

Cryle, P. (2025). *How race theory began to unravel even in its heyday: The (in)discipline of physical anthropology at the time of the Franco-Prussian War (1870–71)* [Manuscript submitted for publication]. Faculty of Humanities, Arts and Social Sciences, The University of Queensland.

Darwin, C. (1871). *The descent of man and selection in relation to sex* (2 vols.). Murray.

de Gobineau, A. (1853–1855). *Essai sur l'inégalité des races humaines* [Essay on the Inequality of Human Races]. Didot.

de Lapouge, G. V. (1899). *L'Aryen, son rôle social* [The Aryan, his social role]. A. Fontemoing.

Gould, S. J. (1987). Petrus Camper's angle. *Natural History, 96*(7), 12–18.

Gould, S. J. (1994). The geometer of race. *Discover Magazine, 15*(11), 64–69.

Gould, S. J. (1996). *The mismeasure of man* (2nd ed.). Norton.

Jackson, J. P., & Weidman, N. M. (2006). *Race, racism, and science: Social impact and interaction*. Rutgers University Press.

Knox, R. (1862). *The races of men*. Henry Renshaw.

La Vergata, A. (2009). *Colpa di Darwin? Razzismo, eugenetica, guerra e altri mali [Darwin's fault? Racism, eugenics, war, and other evils]*. UTET.

Le Bon, G. (1879). Recherches anatomiques et mathématiques su les lois des variations du volume du cerveau et sur leurs relations avec l'intelligence. *Revue d'Anthropologie, 2*(2), 27–104.

Said, E. (1978). *Orientalism*. Pantheon Books.

Sighele, S. (1891). *La folla delinquente* [The delinquent crowd]. Fratelli Bocca.

Taguieff, P.-A. (2005). Racisme aryaniste, socialisme et eugénisme chez Georges Vacher de Lapouge (1854–1936) [Aryanist racism, socialism and eugenics

in Georges Vacher de Lapouge (1854–1936)]. *Revue d'Histoire de la Shoah, 183*(2), 69–134.
Van Ginneken, J. (1992). *Crowds, psychology, and politics, 1871–1899*. Cambridge University Press.

CHAPTER 4

Italy: The North, the South, and the Racism

Recent historical studies have shown that, between the nineteenth and twentieth centuries, at least two foundations of scientific racism were laid:

1. The idea that physical and psychological traits were stable and immutable, necessarily passing from one generation to the next; and
2. the belief that there were different human races arranged in a hierarchical order and that, among these, the northern European race was superior to the others (Nordicism).

Among other things, scientists sought to find a correspondence between physical and psychological traits that would justify hierarchical differences, with the top of the hierarchy mainly comprising individuals from the north (Jackson & Weidman, 2006, pp. 101–109).

This latter opinion, that the populations of the Mediterranean basin were less advanced than those of Anglo-Saxon origin, has persisted to the present day, taking root in Italy as well. After all, Italy is located between Northern Europe and Africa and lent itself well to racist ideologies. In the second half of the nineteenth century, a debate emerged regarding the alleged inferiority of southern Italians. Italian positivist scientists were then interested in the populations of the South in order to help solve the social problems that had emerged with the unification of Italy (illiteracy, poverty, crime, industrial, and infrastructural underdevelopment).

R. Foschi, *Racisms*, https://doi.org/10.1007/978-3-032-22962-5_4

To this end, scholars compared the physical and behavioral characteristics of southerners, motivated by the differences between economically more and less developed areas. These studies reinforced the idea of the possible existence of a southern race with negative characteristics that differed from those of northerners. It was Cesare Lombroso and some of his followers, in particular, who considered the inhabitants of the south to be representatives of an inferior race. Moreover, this negative reputation followed southerners as far as the United States. There, the myth of two Italies spread. The north and the south, including the islands, were profoundly different in terms of the physical and psychological nature of their populations and their civilization. According to a popular stereotype, southerners were perceived as individualistic, anti-social, rebellious, apathetic, and criminally inclined, incapable of adapting to liberal society.[1]

Scholars who addressed the southern question from an anthropological and psychological perspective presented different and sometimes conflicting opinions. As highlighted by Cimino and Foschi (2014), some scholars attributed greater importance to the genetic factors of populations, i.e., hereditary biological and mental characteristics (hereditarists), and those who, on the contrary, gave greater importance to historical and environmental factors, as well as social and cultural factors (environmentalists). The southerners we will discuss shortly belonged to the latter group.

As for the hereditarists, however, it can be said that there were at least two predominant, often competing, schools of thought among them. On the one hand, some believed that physical and mental characteristics were determined by a stable genetic heritage, while admitting changes only through sexual selection, i.e., through the crossing of genes between pairs of parents (innatists). This tendency sought to discover signs of atavism or degeneration and prevent them from perpetuating, proposing various "hygienic" measures that ranged from isolating carriers of such characteristics to planning their elimination. Degenerative theories, also linked to alleged racial differences, thus spread like wildfire in nineteenth-century psychiatric culture. They almost became the primary paradigm for explaining madness (Pick, 1989). Innatists considered environmental causes irrelevant; these scholars resolutely emphasized fixity rather than transformation. On the other hand, there were evolutionists

[1] This image was amplified and propagated, particularly right after unification, by newspapers of the Kingdom of Italy reporting on brigandage.

who, while accepting the principle of heredity, believed that it could not be so rigid and automatic. It was thought that heredity was a mixture of the characteristics of the parents and that the environment selected the most suitable characteristics generation after generation through a non-mechanical process. Furthermore, there was a widespread belief that every characteristic could be transformed across generations, according to Lamarck's principle of the inheritance of acquired characteristics (transformism). For these scholars, genetic transformation was possible not only because of the mixing of populations, but also because of environmental factors—such as exchanges between different cultures and, above all, education—which directly modified the genetic heritage. In this sense, this nineteenth- and twentieth-century approach could be defined as neo-Lamarckian.[2] There were, therefore, researchers who were more or less open to Lamarckism or resolutely opposed to it. Even Spencer retained clear elements of Lamarckism, according to which traits could be acquired through exercise and the environment could transform individual characteristics (La Vergata, 2009, pp. 49–51). Transformism was important precisely for the sciences of the mind, which were essentially based on the idea of the possibility of change (Foschi, 2020). However, the different points of view were intertwined among the various researchers, with some showing more open conceptions of transformism and others being more conservative and fatalistic. This influenced the racism of both groups.

As mentioned, Lombroso had strenuously supported the thesis that the tendency to commit crime was fundamentally an innate character trait marked by primitivism (atavism). Although there were scattered traces of environmentalism even in his writings—for example, in his theory of the occasional criminal or in his theory of "meteors"[3]—Lombroso built his profile as a scholar as that of a staunch innatist who collected physical and psychological signs in support of atavism; it was no coincidence that, in

[2] Although Gregor Mendel (1822–1884) first demonstrated hereditary regularities generation to generation, favoring innatism, the gene-environment relationship was considered an ineluctable component of evolutionary mechanisms. As seen in the history of science, heredity models have alternated between more innate and environmentalist perspectives, such as in contemporary epigenetics, where environmental impact on genes resembles neo-Lamarckian hypotheses.

[3] Lombroso wrote a book in which he acknowledged the influence of natural phenomena such as meteors or geographical features on human thought (Lombroso, 1878). In his work *In Calabria*, Lombroso also attributed the backwardness he observed in southern Italy to environmental causes (Lombroso, 1898).

searching for such signs, he believed he found them in individuals from southern Italy.

In fact, we recall that Lombroso first sought confirmation of his criminological theory in the skull of the alleged Calabrian outlaw Giuseppe Villella, a poor farm laborer, very far from the profile of the born criminal, who had stolen food out of necessity and was therefore harshly condemned and died in prison from illness. He was one of many others born and raised in a tough place and time. Analyzing his skull, he thought he had identified, as a distinctive physical characteristic of the born criminal, a particular "occipital dimple" on the back of the skull, inside the skull, which he considered hereditary and interpreted as a sign of evolutionary backwardness. Villella's own biography disproved Lombroso's discovery of the occipital dimple as a sure sign of the born criminal, mainly because even if the skull studied by Lombroso did indeed belong to Villella, the Calabrian farmhand certainly did not have the criminal record of a hardened criminal, as Lombroso's popular theory subsequently claimed (Milicia, 2014).

Another famous criminal studied by Lombroso was the Calabrian soldier Salvatore Misdea, guilty of killing and wounding numerous fellow soldiers from the north, with whom he had argued about their respective origins. Based on Lombroso's expert opinion (in his defense!), Misdea was sentenced to death, and Lombroso, in this case, broadened the range of evidence of atavism, diagnosing the defendant as a born criminal, partly because he had epilepsy and other family pathologies. This alleged evidence of atavism, therefore, made Misdea a born criminal, irredeemable, and dangerous. Misdea's case was particularly significant both for the gradual introduction of supposed epilepsy in place of the occipital dimple as evidence of innate and biologically determined criminality. It was also significant because Lombroso's expert opinion ultimately contributed to Misdea's death sentence for "his own good" and for the good of society, which had to defend itself from what was considered an irredeemable and uncontrollable savage due to his atavistic and racial origins. There was no other way for Misdea but death, which would put an end to his unfortunate suffering. Misdea thus represented one of the first victims of the ideology of "lives unworthy of being lived" that we find in Nazi eugenics.

Misdea was depicted in a drawing by Lombroso in a markedly animalistic way, with a deformed skull representing an abnormal shape that corresponded to the (im)morality of the offender (Fig. 4.1). This image

Fig. 4.1 Salvatore Misdea portrayed on the cover of Lombroso and Bianchi 1884

did not correspond to other contemporary depictions of Misdea, in which he appears as a normally proportioned person (Romeo & Cefalì, 2015). Cesare Lombroso and Leonardo Bianchi (1848–1927) described him as follows in their expert report:

> The dolichocephalic shape of the skull is typical of Calabrians, but what distinguishes his is the asymmetry in favor of the right frontal lobe, which is less prevalent in healthy men, and above all the reduction in size of the frontal area at the sides with the protrusion of the orbital angle, which are characteristics of microcephalics and lower animals [...]
> Exaggerated craniofacial asymmetry with facial torsion on the skull is even more important because it is one of the most common characteristics of people with epilepsy [...]
> Slight strabismus may indicate innervation defects, as suggested by dullness, inequality, and lateralism of tactile sensitivity and, even more so, of pain sensitivity, as I have just shown [...]

> Congenital dental aplasia, the absence of the two incisors, is important; a characteristic that is not even found in savages; as with the frontal occipital dimple, which I found so frequent in them, we must search for its atavism in rodents [...]
> Add to this the diastema, the distance between the canines and incisors, which is characteristic of many African races and many monkeys [...] Finally, add the overlapping of the incisors of the lower jaw, so frequent in degenerates.
> At first, it makes one smile that importance is given to the displacement of a bone, to the congenital lack of two teeth - as if one thought with one's teeth - but the smile ceases (at least in the thinker) to give way to profound reflections in those who remember how certain very serious, congenital mental alienations, imbecility, cretinism, moral insanity—which is the same as congenital delinquency—are accompanied by certain special features that allude to arrested development, which gradually reproduce the various states of the inferior races, of which eurignathism—frontal sinuses—prognathism—diastema—the lack of two incisors—are actual passages or characteristics. (Lombroso & Bianchi, 1884, pp. 32–35)

Lombroso and Bianchi professed to be Darwinists and, based on science, believed they could prove that Misdea was irresponsible, but not therefore unpunishable. For them, punishment had to be aimed at defending society from the born criminal, and the death penalty would be necessary to "disassimilate" Misdea, remove him from society, and prevent him from passing on his inclinations to his descendants:

> When there are criminals, madmen even, but so dangerous that we are sure they will cost the lives of other men, honest men at that, and such is the case with Misdea, who promised to carry out new massacres, we must assimilate them: if imprisonment is not enough, we must also resort to death, which would give the most complete assimilation with the addition of two less damages: the expense and the impossibility of recidivism, with the added advantage of intimidation, which often deters occasional criminals and, while not to be sought *directly*, cannot be rejected when obtained *indirectly through elimination*. (Lombroso & Bianchi, 1884, p. 112)

Some argue that Lombroso's work, as that of a Lamarckian socialist, should be considered in a less innate sense. However, in the case of Misdea, as in many passages of his writings on the born criminal, Lombroso demonstrates a hard-line innatist attitude, in which biological and racial factors prevail over environmental conditions, which are relegated to the background and considered insignificant. Misdea and those like him cannot be redeemed; they are not guilty because their biology determines their behavior, but they are increasingly seen as criminals from whom society must defend itself, despite their irresponsibility. The "new" legal principles that the Italian positivist criminal school sought to institutionalize were based on determinism and biology, employing a logic that would later be adopted by fascist criminology. Pseudoscientific eliminationism, as we shall see, also gained ground mainly because of those who, like the Nazis, justified their eugenic elimination plans on the grounds that citizens with what they considered mental or social pathologies were harmful to themselves and society. These unaware people were considered dangerous and undesirable. Not only were the victims eliminated, but they were even persuaded that their elimination was for their own good.

One of the last famous cases of southerners examined by Lombroso was that of the Calabrian bandit Giuseppe Musolino, who later became something of a legend for the people of the south, but for Lombroso, just a criminal predisposed to crime because he also had epilepsy and came from a population that, due to Greek and Albanian influence, had become barbaric and showed little moral sense (Lombroso, 1902, p. 513). The considerations on Musolino represented a further refinement of research into the innate biological causes of criminal behavior, and epilepsy became the distinctive (almost exclusive) proof of delinquency determined by innate causes. However, Lombroso's idea of epilepsy and epileptic familiarity was very generic, a cause of criminality destined to be criticized, as was the previous idea of the occipital dimple. Following this line of thinking, Lombroso's heirs were consistent with the theory of atavism. In the positivist school of criminal law, whose leading representative was Enrico Ferri (1856–1929), the right of society to imprison criminals due to their dangerousness, attributed to an innate tendency toward crime, was upheld. In this way, this "new" school affirmed the principle that imprisonment was not so much an act of expiation as a means of protecting society from the danger posed by criminals.

In his work, Ferri argued that southern criminals (Neapolitans, Calabrians, and Sicilians) had smaller skulls than normal people. More generally, believing it possible to identify racial varieties from different European regions based on specific skull characteristics typical of superior or inferior populations, he classified southerners as inferior and prone to criminality (Ferri, 1895). This southern paradigm applied to all peoples of southern Europe. Although Ferri fell into numerous contradictions, throughout the book, he argued that southerners were murderers because of their physical constitution and genetics, which were mixed with those of Africans, while other crimes, judged less severely, such as theft, were more typical of those with skulls characteristic of northern regions. About Spain, citing some local statistical and economic data, Ferri stated:

> In the southern provinces, however, from the beginning of the eighth century onwards, there was a complete invasion and domination by the Saracens (Arabs and Berbers), for whom "through continuous interbreeding over the centuries, the original type has changed, as evidenced by the type of inhabitants in the southern districts ... In which the population had become so 'African' that they differed from Africans more in their religious practices than in their skin color ...; And after the Arabs, no other important ethnic group overlapped with them ... Moreover, although the Inquisition expelled from the Kingdom or enslaved perhaps a million Moors, this does not detract from the fact that the lower strata of the population were already 'Arabized', both in their features and in their character, due to a certain fatalism that was entirely Muslim."
>
> Thus, Reclus says, "the truly African Spain, in terms of its general appearance, its flora and fauna, as well as its populations, begins only in the treeless highlands of the interior and especially on the Mediterranean shores." And Mougeolle, among others, observes that on the Iberian Peninsula, "the frontier of Arab civilization can be fairly well represented by a diagonal line between Lisbon and Barcelona. Everything south of this line belonged to the Moors."
>
> However, a glance at cartogram 42 shows that the highest incidence of violent crime is south of this diagonal line, in the more Arab provinces of Spain. In contrast, cartogram 43 clearly shows that theft is more frequent north of this diagonal line. (pp. 288–289)

The general idea of positivists such as Lombroso and Ferri was, therefore, that inferiority stemmed from the genetic influence of the Phoenicians, Arabs, and "Moors," who were considered inferior to the Germans and Celts, who had instead hybridized with the inhabitants of the north.

In these pseudoscientific narratives, populations were explicitly treated as distinct racial groups. For these scholars, it was entirely self-evident that there were multiple human races characterized by distinct physical characteristics, corresponding to varying intellectual and moral levels. For these positivists, the populations of the north, considered more intelligent and advanced, exhibited more "evolved" criminal acts (fraud and robbery) than the crimes of the south (murder).

Along the same lines, Gaspare Virgilio (1836–1908), a Lombrosian and director of the Aversa Mental Hospital, searched for predispositions to crime and madness, for hereditary physical and mental traits, which he found, for example, in Giovanni Passannante (1849–1910), a poor cook from Basilicata who, in 1878, was responsible for a naive attempt to kill King Umberto I with a penknife. During the trial, both Lombroso and Virgilio, contrary to the more moderate positions of other experts (such as Tamburini), insisted on considering Passannante a criminal and a madman due to obvious family defects. After the trial and conviction, Virgilio had his brother committed to an asylum, and the entire Passannante family was labeled degenerate and socially dangerous, forcing some members to change their surname. Even Passannante's village was forced to change its name from Salvia di Lucania to Savoia di Lucania to atone for the sins of its unfortunate citizen (Galzerano, 1997).

Even more "racist" were the descriptions and evaluations of the southern population by a Sicilian follower of Lombroso, Alfredo Niceforo (1876–1960), the true inventor of the "two Italies." In his 1898 book *L'Italia Barbara Contemporanea* (Contemporary Barbaric Italy), Niceforo clearly expressed his conception of the existence of a northern Italy and a southern Italy, inhabited by distinct races: in the north, the evolved Aryans, Germanic peoples of Eurasian origin; in the south, the backward Latins, Mediterranean peoples of Euro-African origin. For Niceforo, the Aryans were more modern and refined, while the Latins were more animalistic and primitive; even crime was more refined in the former group (fraud) and primitive in the latter (murder, robbery, kidnapping). He revisited these theories in his 1901 book *Italiani del Nord e Italiani del Sud* (Northern Italians and Southern Italians), in which, following criticism of his early writings, he added some environmental considerations, while reiterating the priority of innate factors. Lombroso, Ferri, Niceforo, and Virgilio thus represented a school of naturalists who believed in the theory of atavism and interpreted Darwinism in an extremely innatist way, only marginally oriented toward transformism.

The case of one of the most illustrious Italian anthropologists of the nineteenth century, the Sicilian Giuseppe Sergi (1841–1936), was different. He was also a staunch supporter of hereditary and evolutionary positivism, a naturalist who was politically involved in the activities of the *International Association of Free Thought*, and a promoter of the libertarian pedagogy of Francisco Ferrer (1859–1909) (Cicciola & Foschi, 2017)—remembered for his work in craniology and for having also indicated the presence in Italy of two different racial groups: the Aryans and the Italic-Mediterraneans. In his opinion, however, these were very mixed and could be distinguished mainly by their psychological characteristics (the social sentiment of the Aryans and the individual sentiment of the Italic-Mediterraneans) and the type of civilization (industrial society versus peasant society). In addition, Sergi, one of the founders of experimental psychology and education in Italy, defended a concept of educability of a transformative nature which, during the twentieth century, influenced his students Sante De Sanctis (1862–1935) and Maria Montessori (1870–1952). Ultimately, Sergi believed that character traits could be modified and passed down from generation to generation, both through the mixing of different genetic backgrounds and through the influence of education, especially "scientific" education. On a descriptive level, Sergi was also convinced of the existence of psycho-anthropological differences between the north and south, but he considered Niceforo's analysis too harsh and crude. He believed in the possibility of classifying people into distinct groups based on craniological characteristics, but argued that Italian populations were a mixture that was difficult to untangle. Moreover, he claimed that psychological differences were primarily determined by social history, i.e., by recent historical events which, in the case of southerners, had not been able to reawaken the potential of their glorious past. In other words, he did not believe so much in racial differences as in psychological and cultural differences, which, even if they had a hereditary basis, remained, in any case, modifiable. The task of the environment was to stimulate the expression of hereditary potentialities that, in specific historical periods, had remained dormant.

Furthermore, during World War I, Sergi supported the alleged natural superiority of southern individualism over German gregariousness. In fact, he believed that the individualism typical of southerners was capable of achieving ingenious results and had led to the success of the Risorgimento against the gregariousness of the Austrian oppressors. Thus, while Lombrosian innatism believed that the peoples of the Mediterranean basin (Greeks, Albanians, Arabs) had contaminated and weakened the population of southern Italy, Sergi's transformist model believed that southerners possessed favorable natural characteristics that only needed to be stimulated by the environment. Hence, the search for solutions, such as the mixing of different populations, imitation, and education (Sergi 1916a, 1916b). Sergi's reversal of the position that southerners were better than northerners was useful in justifying anti-German ideology during the First World War. In France, too, rivalry with Germany prompted French positivists to try to prove, through physical anthropology, that Germans were inferior, using *ad hoc* criteria (Van Ginneken, 1992). The Calabrian Pasquale Rossi, an original Italian crowd theorist, also observed that the negative differences between southerners and northerners were determined by the socio-economic problems of the south, due to contingent historical and political reasons, which had stifled the population's best potential and instead favored the emergence of less favorable predispositions.[4]

In this way, both Rossi and Sergi seemed to conceptualize race as a malleable and modifiable notion, in the sense that the distinctive traits of a racial group could be transformed, even biologically, through pedagogical and socio-economic interventions.

Toward the end of the nineteenth century, lively criticism developed against psycho-anthropologists, both innatists and transformists, led by a group of scholars from diverse backgrounds who formed the so-called Meridionalisti (Southerners). Among them, a historically important figure was the Sicilian Napoleone Colajanni (1847–1921), a democrat and republican, politician, statistician, and sociologist of great value. In his 1894 book *La delinquenza in Sicilia e le sue cause* (Delinquency in Sicily and its causes), Colajanni completely shifted the focus of the discussion on delinquency in the South, which, in his opinion, stemmed from political and economic factors, such as the exploitation of the poorest classes in Sicily. He was convinced that the differences between north and south were due to the different regimes of impoverishment in the south, of which the Bourbons and the Savoy monarchy represented only the final part of a complex history. Colajanni contextualized the southern question within a broader economic crisis; he also harshly criticized Lombroso and his followers, in particular his fellow countryman, Niceforo. In a 1898 short essay entitled *Per la razza maledetta* (For the accursed race), Colajanni highlighted all the inconsistencies in Niceforo's simplistic thinking, which was based on a series of contradictory craniological data. Instead,

[4] Contrary to Sighele and Le Bon, Rossi believed the crowd necessarily did not have regressive characteristics, which, among early social psychologists, made him unique. Cf. Rossi (1898).

he used socio-economic statistical indicators, highlighting that illiteracy rates in southern Italy were correlated with delinquency and backwardness. For him, too, therefore, the solution to the southern question lay in educational policies.

Along the same lines, the great Italian southernist Gaetano Salvemini (1873–1957) gave no credence to anthropological studies on racial differences between the north and the south. In his opinion, the notion of race, for the purpose of explaining social phenomena, was only suitable for lazy people with no desire to delve into complex issues. For Salvemini, race was the result of history, not one of its causes. He did not deny the character and behavioral differences between the North and South, but he was fully convinced that they stemmed from the different histories of the Italian regions. In the south, where there was no entrepreneurial class or proletariat, social relations had been conditioned by a feudalism that persisted throughout successive regimes (Salvemini, 1900, 1955).

After the unification of Italy, some Italian scientists described southerners in racial terms as individualistic and rebellious (and therefore potentially criminal and incapable of building a civil society), ultimately characterizing them as more primitive and less evolved. In contrast, northerners were considered more orderly, sociable, and better able to organize themselves, but also more gregarious and submissive to authority.

The southern question had also erupted in Italy at a historical moment when Nordicism was gaining traction in Europe and America, contrasting a hegemonic and victorious Aryan and Anglo-Saxon civilization with a Mediterranean one (Greco-Roman, with Arab influences), now considered decadent. What is more, all theorists of racism were convinced of the superiority of the Nordic race and, starting with Arthur de Gobineau, the issue had become central (de Gobineau, 1853–1855).

Only those like Colajanni, who had cutting-edge sociological expertise, strongly rejected the idea that there could be a northern race and a southern race, considering the differences to be caused by noticeable environmental and socio-economic differences that also created moral inequality. Colajanni analyzed all the contradictions in the classifications of racist positivists, all the measurement and anthropometric inconsistencies, and the theoretical weaknesses of both polygenists and monogenists, arguing quite convincingly that races do not exist and that the word itself is wrong. He preferred to talk about civilizations and nations that had nothing to do with the biology of their inhabitants but instead with culture, economics, and politics:

The stark contradictions recorded throughout history between the anatomical and psychological characteristics of different *races* should have led scientists to exercise greater caution in assigning intellectual or moral characteristics as special, distinctive, or exclusive to one race or another. However, this has not been the case …
In truth, it is not easy to bring order and clarity to this subject. It is worth noting that pure races do not exist from a strictly anthropological perspective. We lack authentic documents to describe the characteristics of primitive peoples …
Referring to contemporary peoples and nations, it is even more difficult—indeed impossible—to distribute characteristics to races: Italy is home to brachycephalics and dolichocephalics; in Germany and also in Prussia, which today represents the preeminent and leading part of the empire, Finns, Slavs, and Saxons are inextricably mixed; in France, representatives of *homo europaeus*, *alpinus*, and *meridionalis* rub shoulders; in Austria-Hungary, all the sub-races further complicate the problem; and worse still in the Russian Empire, where a great deal of Mongolian blood flows in the veins of the boyar or mujick. Finally, the confusion is compounded by the varying meanings assigned to certain words, which are used so loosely. For Ammon, Closson, Lapouge, Woltmann, and several others, the Aryans are tall, blond dolichocephalics; and for Ammon himself, the Aryans are the Germans. For Sergi, and for many others, the Aryans are the brachycephalic descendants of *Homo alpinus* scattered throughout central France, northern Italy, southern Germany, Austria, and Russia!
Nor is the repetition of these contrasts a whim or a desire for pleasure; rather, it serves to show how much trust such attributions deserve when we see that among the leading anthropologists, the superior psychic characteristics that others grant some attribute to dolichocephalics to brachycephalics…
Therefore, those who, when speaking of characteristics, refer to national types that have been formed over the centuries through the fusion of races, acclimatization, progressive or regressive selection, and the hereditary transmission of acquired characteristics, are more prudent.
These products have been influenced by the set of intellectual and moral factors that constitute a civilization. This is why the products of the fusion of various races in France, Italy, and Spain have been improperly referred to as Latin *peoples*, when they could only be nations with a Latin *civilization*. The distinction between *race* and *nation* is necessary because one is distinguished by anatomical characteristics and the other by psychological characteristics. When referring to the latter, therefore, one must always refer to the people, the *nation*, and not the *race*. (Colajanni, 1903, pp. 30–32)

Colajanni was one of those scholars who, many years before Nazi-Fascist racism, had pointed out that race scientists were actually engaged in pseudoscience, that their notions were contradictory, and that racist descriptions and classifications were fanciful. The biology that underpinned racism, therefore, seemed very weak even at the time, yet for a long time and until recently, it continued to enchant those who wanted to find a false scientific basis for prejudice (Teti, 2011). In his books, Colajanni demolished the prejudices of the positivists one by one, demonstrating with data that the socio-psychological reality of the peoples studied, in addition to not being racializable, was often the opposite of what the racists wrote. Although anti-racists presented convincing data to support their hypotheses, racism continued to take root based on what seemed obvious, namely, the phenomenal differences between populations, considered to be the biological root of racial hierarchies.

Today, at least in Italy, no serious scholar goes so far as to hypothesize apparent racial differences between the north and south; yet, in the face of economic differences, many believe it is necessary to find the causes, especially cultural ones. After World War II, starting with the studies of the American Edward C. Banfield (1916–1999), the traditional personality trait of southern Italian individualism as a cause of backwardness was transformed into that of *amoral familism*, explicitly considered the opposite of the tendency to associate, which, according to Anglo-Saxon tradition, is the basis of an advanced and democratic society (Banfield, 1958; see also Foschi & Lauriola, 2016).

One of the most renowned contemporary sociologists, Robert D. Putnam, starting from Banfield's idea, argued instead for an interpretation similar to that of the southerners, concluding that it was historical circumstances and, in particular, persistent feudalism that produced socio-cultural differences and a subservient mentality due to backwardness. For Putnam, the north, unlike the south, evolved into a different culture, derived from city-states, where forms of participatory government already existed in the Middle Ages (Putnam, 1993).

The international debate on psychological and cultural differences between North and South is therefore still alive. In this debate, biological and racial arguments have been definitively overcome. In 2010, British psychologist Richard Lynn, who, as we shall see, belongs to a circle of hard-line psychometricians that we will discuss in the following pages, published some research in accredited psychological journals arguing that the demographic, economic, and intelligence differences between northern and southern Italy were caused by genetic factors. For Lynn, the biological differences of northern Italians were the cause of greater intelligence and wealth (Lynn, 2010). Moreover, Lynn's studies are part of a broader research project aimed at demonstrating, worldwide, genetic differences between southern and northern populations that would result in IQs leading to different socio-economic development. Although overwhelming evidence exists against scientific racism, Lynn himself still uses racial concepts, despite the numerous and well-founded criticisms he has received, especially in recent years, which we will discuss in the final chapters.

In conclusion, the tendency to attribute maximum importance to predisposing and hereditary factors to justify differences between north and south has not disappeared; just as there is still a casual use of the notion of race, especially in Anglo-Saxon countries. The issue of potential racial differences between populations of different nations, or between groups of people within the same territory, continues to draw the attention of researchers. Moreover, such studies often operate independently of contemporary genetics, which, as we have seen, has rejected the notion of race as unsuitable for human biology.

References

Banfield, E. C. (1958). *The moral basis of a backward society*. Free Press.

Cicciola, E., & Foschi, R. (2017). Giuseppe Sergi tra pensiero positivista e impegno politico [Giuseppe Sergi between positivist thought and political engagement]. *Physis, LII*, 169–182.

Cimino, G., & Foschi, R. (2014). Northerners versus southerners: Italian anthropology and psychology faced with the "southern question". *History of Psychology, 17*(4), 282–295. https://doi.org/10.1037/a0036547

Colajanni, N. (1894). *La delinquenza in Sicilia e le sue cause* [Delinquency in Sicily and its causes]. Sandron.

Colajanni, N. (1898). *Per la razza maledetta* [For the accursed race]. Sandron.

Colajanni, N. (1903). *Razze inferiori e razze superiori o Latini e Anglo-Sassoni* [Inferior races and superior races or Latins and Anglo-Saxons]. Rivista Popolare.

de Gobineau, A. (1853–1855). *Essai sur l'inégalité des races humaines* [Essay on the inequality of human races]. Didot.

Ferri, E. (1895). *L'omicidio nell'Antropologia Criminale* [Murder in criminal anthropology]. Bocca.

Foschi, R. (2020). *Storia della psicologia e della mente* [History of psychology and the mind]. Mondadori.
Foschi, R., & Lauriola, M. (2016). Do amoral familism and political distrust really affect north–south differences in Italy? *Journal of Cross-Cultural Psychology, 47*, 751–764.
Galzerano, G. (1997). *Giovanni Passannante. La vita, l'attentato, il processo, la condanna a morte, la grazia "regale" e gli anni di galera del cuoco lucano che nel 1878 ruppe l'incantesimo monarchico* [Giovanni Passannante: Life, the assassination attempt, the trial, the death sentence, the "royal" pardon and the prison years of the Lucanian cook who in 1878 broke the monarchical spell]. Galzerano Editore.
Jackson, J. P., & Weidman, N. M. (2006). *Race, racism, and science: Social impact and interaction*. Rutgers University Press.
La Vergata, A. (2009). *Colpa di Darwin? Razzismo, eugenetica, guerra e altri mali* [Darwin's fault? Racism, eugenics, war, and other evils]. UTET.
Lombroso, C. (1878). *Pensiero e meteore* [Thoughts and meteors]. Dumolard.
Lombroso, C. (1898). *In Calabria*. Giannotta.
Lombroso, C. (1902). L'ultimo brigante. Giuseppe Musolino [The last bandit: Giuseppe Musolino]. *Nuova Antologia, 181*, 508–516.
Lombroso, C., & Bianchi, L. (1884). *Misdea e la nuova scuola penale* [Misdea and the new criminal school]. Bocca.
Lynn, R. (2010). In Italy, North–South differences in IQ predict differences in income, education, infant mortality, stature, and literacy. *Intelligence, 38*, 93–100.
Milicia, M. T. (2014). *Lombroso e il brigante* [Lombroso and the brigand]. Salerno editrice.
Niceforo, A. (1898). *L'Italia barbara contemporanea* [Contemporary barbaric Italy]. Sandron.
Niceforo, A. (1901). *Italiani del Nord e Italiani del Sud* [Northern Italians and Southern Italians]. Bocca.
Pick, D. (1989). *Faces of degeneration: A European disorder, c. 1848–1918*. Cambridge University Press.
Putnam, R. D. (1993). *Making democracy work*. Princeton University Press.
Romeo, D., & Cefalì, F. (2015). *Salvatore Misdea. 1884: follia criminale o determinazione di un soldato del sud Italia?* [Salvatore Misdea. 1884: Criminal madness or determination of a southern Italian soldier?]. Gigliotti.
Rossi, P. (1898). *L'animo della folla* [The soul of the crowd]. Riccio.
Salvemini, G. (1900). *Risposta* [Reply]. In A. Renda (Ed.), *La questione meridionale. Inchiesta* [The southern question. Inquiry] (pp. 38–42). Sandron.
Salvemini, G. (1955). *Scritti sulla questione meridionale (1896–1955)* [Writings on the Southern Question (1896–1955)]. Einaudi.

Sergi, G. (1916a). La cultura Germanica [The Germanic culture]. *Nuovo convito, 1*, 7–9.

Sergi, G. (1916b). Gli istinti gregari nell'uomo e il popolo tedesco [The gregarious instincts in man and the German people]. *Nuovo convito, 1*, 151–154.

Teti, V. (2011). *La razza maledetta. Origini del pregiudizio antimeridionale* [The cursed race: Origins of anti-Southern prejudice] (2nd ed.). Manifestolibri.

Van Ginneken, J. (1992). *Crowds, psychology, and politics, 1871–1899*. Cambridge University Press.

CHAPTER 5

Eugenics

Eugenics was an invention of English psychologists that was successful in the Anglo-Saxon world and subsequently spread throughout the West. As a result, many historians of science have studied it, and it has become a well-known topic among specialists (see Kevles, 1985 on eugenics). Closely related to scientific racism, eugenics was a discipline developed by Francis Galton (1822–1911), Darwin's cousin, in the second half of the nineteenth century. Its aim was improving humanity by actively selecting individuals who carried hereditary characteristics considered most adaptive (Galton, 1883). These characteristics were identified in the intellectual and moral qualities of individuals. Eugenics was therefore a scientific attempt to do a "service" to humanity by promoting or selecting the characteristics considered best in both individuals and groups of individuals. Several evolutionists and politicians promoted eugenics as a means to select desirable characteristics in a population, with the goal of defending society through genetic intervention. This led to the emergence of positive eugenics, which aimed to promote specific characteristics in the population, for example, by promoting demographic policies. There was also negative eugenics, which essentially acted by controlling the possibility of transmitting the genetic makeup of specific individuals, limiting marriages, sterilizing individuals or families burdened by various psychopathological diagnoses, or eliminating entire groups of individuals. From this perspective, eugenics and racism were often

R. Foschi, *Racisms*, https://doi.org/10.1007/978-3-032-22962-5_5

a dangerous combination involving institutions, states, scientists, politicians, and philosophers. If certain races were inferior and needed to be controlled or eliminated, eugenics became the secular arm of racism. The history of eugenics is therefore quite long and complex, and has been associated with various authoritarian regimes. However, it is also worth noting that eugenics was a characteristic phenomenon of democracies. Many American states instituted eugenics laws that remained in force until a few decades ago, legislation that favored the sterilization of people with varying degrees of psychopathology, often African Americans. Similarly, in Northern Europe, and specifically in Sweden, until the 1970s, there were eugenic sterilization laws targeting individuals with mental disorders of varying severity, even promoted by social democratic governments to contain public spending of national health systems (Broberg & Roll-Hansen, 1996; Colla, 2000).

As highlighted by La Vergata (2009), Galtonian eugenics was essentially a utopian project with the characteristics of a religion.[1] Around 1911, Galton even wrote a novel, which was actually more dystopian than utopian and remained unpublished until a few years ago, about a future ideal society based on eugenics, Kantsaywhere (I-don't-know-where or Kant-indicates-where), a place based on evaluation where citizens, "tested" in a college that ruled the city, were awarded diplomas with scores for various psychological, aesthetic, and anthropometric abilities and qualities; depending on the score achieved, inhabitants could reproduce gradually, from the highest degree of freedom for the best, to the worst on the scale of values who had to live in abstinence and forced labor in colonies (Galton, 1911/2001).[2]

Eugenics was therefore a peculiar field, conceived by Galton based on his particular worldview, which was biological and psychological, a conception whereby tests, numbers, psychometrics, and increasingly refined measurements could improve the fate of society. It was a worldview that lent itself to being closely associated with racism. Eugenics

[1] Moreover, Galton's utopian conception was likely influenced by the progressive Anglo-Saxon Masonic culture. Erasmus Darwin (1731–1802), the family patriarch, was a naturalist and abolitionist, grandfather to Francis Galton and Charles Darwin, and a leading Anglo-Saxon Mason. Many of his children were Masons, and Galton himself was initiated at a young age in an academic lodge. His youthful assimilation of progressive ideals evolved into an unshakable faith in societal improvement via statistics and eugenics.

[2] Note that Galton's family was large: Erasmus fathered fourteen children. See the following pages and cf. Ray (1983).

itself could take on various forms and interpretations: there could be eugenic neo-Lamarckism, eugenic socialism, eugenic hygienism, eugenic Catholicism, and even some eugenic feminists such as Sybil Gotto (Sybil Neville-Rolfe) (1885–1955), who promoted eugenics for hygienic and social purposes as a progressive political commitment (Ray, 1983).

Eugenics is therefore inextricably linked to the name of Galton and his students, who, as is well known, were also important promoters of the birth of scientific psychology and, specifically, of the so-called Anglo-Saxon psychometric tradition. Eugenics should therefore also be fully included in the history of psychology, as it has influenced its study and methods. In particular, the psychology of personality and individual differences has its roots in the Galtonian tradition. However, eugenics is not yet a particularly prominent topic in the historical training of psychologists and is not universally included in psychology history textbooks.[3]

Eugenics was also a phenomenon derived from English scientific psychology. Starting from a particular empiricist attitude, Anglo-Saxon philosophical psychology distinguished itself as a movement that reformed the humanities and social sciences on a broad scale, dealing with psychological issues alongside political, logical, and economic problems. Anglo-Saxons involved in psychology can thus also be considered the fathers of liberalism and illustrious founders of modern law.[4]

In the English context, Charles Darwin was of fundamental importance to the history of psychology. First, he proposed that human evolution should be treated in the same manner as the evolution of other animal

[3] Only recently have some critical authors explored the eugenic aspects of twentieth-century psychology. Nikolas Rose, in his *Psychological Complex*, rightly identified eugenics as a pillar of English psychology; Leahey, in *A History of Psychology: Main Currents in Psychological Thought*, pointed to eugenics as a source of American functionalism, extending to an eugenic approach to a differential intelligence study aimed at protecting American society from immigration of "unsuitable" Europeans seen as a threat to the U.S. genetic pool. Among younger historians, the issue is more frequently addressed, especially in relation to Galtonian innatism and racism. However, even today, textbooks often lack specific sections that connect general and personality psychology with eugenics history. Cf. Leahey (1980), Rose (1985), and Walsh et al. (2014). See also Foschi (2019).

[4] Starting with David Hartley (1705–1757), those interested specifically in psychological topics distinguished themselves by accepting the idea of an independent psychology, though initially this did not equate to full autonomy from philosophy. The attempt by English empiricist philosophers to establish independent psychology had limited success compared to the later influence from evolutionary positivists. Cf. Foschi (2020).

species. In *On the Origin of Species* (1859), he also proposed the fundamental idea that natural selection rewards those inherited characteristics, even in slightly modified form, that are more likely to be adaptive (pp. 60–61). In Darwin's theory, however, the environment played a significant role. Darwin, therefore, remained linked to Lamarckism through his theory of pangenesis, a hypothetical biological mechanism that posited how changes were inherited. He believed that gemmules were inherited sexually from parents to children, that they were affected by parental experience, and could therefore be more or less adaptive in the second generation. The evolutionary process outlined in the theory of pangenesis did not, therefore, reject the idea that exercise could lead to the transmission of characteristics from one generation to the next through gemmules. In that sense, it was influenced by Lamarckian thinking. However, eugenics was successful and became established as an application of biology when an innatist conception of heredity replaced Lamarckism.

Years after Darwin, based on the work of August Weismann (1834–1914) on the transmission of germ cells, the idea that acquired characteristics could be inherited fell into disfavor in mainstream biology, giving rise to a strict form of hereditarianism (hard inheritance vs soft inheritance) (Mayr, 1982, pp. 693–698). This hereditarianism was based on demonstrations of the theory of germplasm continuity, according to which genetic traits were transferred from one generation to the next without any intervention from the environment, and on the natural selection of variations inherent in individuals and contained in their germplasm. The laboratory study of chromosomes and chromatin that was developing in the last decades of the nineteenth century, including by Weismann himself, seemed to have dealt a fatal blow to neo-Lamarckism. Social scientists, psychologists, and psychometricians were divided between those who remained neo-Lamarckian transformists and those who began to take polarized positions whereby all behavior was considered innate. Genetics would thus be considered a sort of deus ex machina with which one could understand humans, predict their behavior, and control it; one could thus also think of managing the masses (bio)politically by integrating psychometrics, biology, and politics, just as in a dystopian world.

The study of the evolution of adaptive characteristics of psychological facts thus became one of the typical elements of evolutionary psychology, but Galton took a decisive step forward. Embracing the radical and innatist conception of heritability, unlike Darwin and independently of Weismann, he emphasized an essentialist idea of individuals, placing them in homogeneous groups based on the statistical average of specific characteristics, from physical (or anthropometric) to elementary psychological ones. In Galton's view, characteristics were considered stable and passed from one generation to the next, and could be either adaptive or maladaptive, desirable or undesirable, inferior or superior. In neo-Lamarckian transformational evolutionism, decisive importance was given to the environment, change, and the exercise of faculties to improve the fate of human nature from one generation to the next. In Galton's interpretation of heritability, importance was attributed to genetic variations due to sexual selection.

According to this biological perspective, Galton initiated his study of hereditary psychological characteristics, making extensive use of statistics. The differential study of individuals based on their means and deviations from the mean became the primary method for understanding which characteristics humans should possess to increase their chances of survival. This also represented the search for the desirable essences of human nature, and deviations from the mean were considered undesirable errors. This gave rise to English psychometrics, a tradition of research that was an alternative to German Wundtian psychology and has continued to this day.[5]

Galton's discourse, however, was not limited to measurements, statistics, and reasoning about the evolution of human behavior. Galton, emphasizing specific characteristics of the evolutionary message such as survival of the fittest and the struggle for existence, vehemently pursued the idea that human selection should be encouraged, transforming selection from natural to artificial, in order to promote new generations of individuals. These individuals would be improved through interventions in families to prevent less gifted or problematic individuals from mating and instead favor the union of those who were desirable according to the standards of Victorian culture. This was the primary objective of eugenics. It was understood as a practice aimed at improving the human race, based on knowledge gained through the measurement of characteristics that were functional for the genetic improvement of populations. The task

[5] As is known, besides the German and English traditions, the other scientific psychology tradition was the French pathological one. Cf. Danziger (1990).

of the behavioral scientist would be to promote those adaptive qualities characteristic of individuals considered superior (positive eugenics) and to hinder those negative characteristics that should instead be socially controlled or even eliminated (negative eugenics). Galton contributed, for example, to the spread of birth control (Galton, 1889; see also Gillham, 2001; Mayr, 1982, p. 623). To pursue his goals, he became a true expert in many scientific disciplines and statistical methods. Much of his work was thus aimed at putting his knowledge and techniques at the service of eugenics. Galton has been called the apostle of quantification,[6] as he disseminated an approach in the human sciences that aimed to count frequencies to highlight how human phenomena fell within the broader domain of natural phenomena. From Galton and his school, psychology acquired the principles of psychometrics. Mental qualities could be categorized by counting the frequencies of physical and psychological facts (transforming quality into quantity). They could also be correlated (a physical or psychological phenomenon could be known insofar as its existence was linked to the simultaneous presence of other physical or psychological phenomena), and could finally be interpreted based on deviations from the average, considered as normality.

Galton thus assimilated the work of the Frenchman Adolphe Quetelet (1796–1874) into the field of psychology. Quetelet had applied statistics to the study of anthropometric measurements of soldiers, noting that height and chest circumference corresponded to the bell curve distribution of natural phenomena, a concept already studied by Carl Friedrich Gauss (1777–1855) in relation to the distribution of measurement errors. This was a probabilistic distribution in which it was possible to represent frequencies on a geometric plane, noting that some spontaneous phenomena (such as measurement errors) clustered with greater frequency around the average measurements. Galton applied the Gaussian or normal curve to anthropometric measurements but also to sensory-motor facts, highlighting that they too could be approximated to the normal curve in the same way as height or chest circumference. With Galton, therefore, psychology was able for the first time to equip itself with a measurement apparatus based on the statistics of normality and abnormality, considering the former as more frequent and adaptive

[6] Famous definition by Stephen Jay Gould (1941–2002). Cf. Gould (1996, p. 107).

and the less frequent facts as anomalous, maladaptive, and essentially pathological (Addeo, 2008).

Normal distribution and frequency measurement became Galton's true obsession. He constructed a sort of methodological ontology capable of justifying, through measurements, everything about humans, from physical anthropometry to sensory, motor, and perceptual phenomena, as well as intelligence, referring to the presumed regularity of normal distribution, even for psychic facts. As Danziger pointed out, Galton's research thus inaugurated a form of experimentalism in which the relationship between experimenters and participants differed significantly from those in German and French psychological traditions (Danziger, 1990; see also Foschi, 2020). Galton's "subjects" were neither university students nor patients. Research in England thus shifted to a level where averages were sought in representative samples of individuals who were important to the experimenter because they were included in the frequency count to calculate the average that represented the benchmark and normality. Individuals were recruited by Galton as if he were providing them with a commercial service. Participants even had to pay the researcher three pence to be catalogued, described, and have their data entered into an aggregate set that actually served the researcher and not the participant. Galton advertised his service with a poster, and participants received a folder with personal measurements that were of no use to them. Instead, that data became the sample used by Galton and his students for their research. To recruit a large number of people, Galton also participated in international exhibitions, such as the International Health Exhibition in London in 1884 (Danziger, 1990, pp. 54–56). In other words, Galton's experiment was based on a type of commercial relationship in which individuals paid for a service that was useful to the researcher for purposes often unknown to the participants.

This marked the beginning of psychological discourse that still uses surveys of hundreds and thousands of individuals, more or less representative of a population, to collect data from which to deduce psychological, but also sociological, dimensions present in different ways in each subject and, above all, in the aggregate constituted by the sample. This data was processed according to probabilistic models, the main one being, as already mentioned, Gauss's normal distribution.

Galton's students were therefore mostly statisticians and psychometricians, all influenced by their teacher's eugenic ideas: among them were

Karl Pearson (1857–1936), who developed indices for measuring correlations and regressions, chi-square, and standard deviation; his student Ronald A. Fisher (1890–1962), a probability theorist and creator of modern methods for hypothesis testing; Charles E. Spearman (1863–1945), psychometrician and theorist of general intelligence; the American James McKeen Cattell (1860–1944), who invented the term mental test and was one of the promoters of the American psychological movement. Galton's followers were numerous, primarily due to the widespread adoption of eugenics in Anglo-Saxon circles. In 1902, Galton and his students founded the journal Biometrika, edited by Pearson and still in existence today, as a means of disseminating the statistical approach to eugenics.

In this context, Sybil Gotto, a feminist who described herself as a eugenicist, together with Galton himself and his students, founded the Eugenics Education Society in England, which was mainly involved in promoting hygiene policies. This society, which later became simply the Eugenics Society, had as its goal birth control, with the idea that this would improve society along with other prophylactic and philanthropic interventions. Early eugenics was, therefore, the emanation of a cultured and wealthy bourgeoisie who naively thought that eugenics was a progressive tool (Ray, 1983), a device whose dangers they did not foresee. These were the society's objectives:

1. Persistently to set forth the national importance of eugenics in order to modify public opinion and create a sense of responsibility, in respect of bringing all matters pertaining to human parenthood under the domination of eugenic ideals.
2. To spread a knowledge of the laws of heredity so far as they are surely known and so far as the knowledge might effect improvement of the race.
3. To further eugenics teaching at home, in the schools, and elsewhere (Hall, 1990, p. 328).

Another aspect of the Galtonian tradition was paradigmatic in this regard: the use of intelligence studies to propose eugenic applications. It seemed apparent that an unfortunate life was due to inferior intelligence. However, Galton, in his laboratory, focused on anthropometric measurements or simple sensorimotor dimensions that allowed him to calculate sample averages of the same phenomena. As is well known, it was essentially Alfred Binet (1857–1911) who created the first comprehensive test that directly measured intelligence, originally intended to enable children with learning difficulties to receive specialized education to improve their condition. Therefore, the test was not designed for eugenic purposes such as birth control for the children of the less intelligent, the selection of immigrants, or access to education. However, when Binet's test arrived in the United States, it immediately became a eugenic tool, and the Galtonian tradition reworked the test according to the methodological canons of Anglo-Saxon psychometrics, distorting the Frenchman's work (Foschi & Cicciola, 2006; Cicciola, 2019).

These psychometricians-eugenicists reduced the measures of the original test to the Intelligence Quotient (IQ):

> They assumed that intelligence was largely inherited, and developed a series of specious arguments confusing cultural differences with innate properties. They believed that inherited IQ scores marked people and groups for an inevitable station in life. And they assumed that average differences between groups were largely the products of heredity, despite manifest and profound variation in quality of life. (Gould, 1996, p. 187)

Although not all Galtonians were openly racist, as Mayr (1982) argues, the use of Galtonian eugenics led to an approach that consequently re-established scientific racism, understood as the study of a group of individuals who share common psychological and physical characteristics and to whom superior or inferior innate qualities are attributed. Members of these racial types (or characterologies), as is well known, were discriminated against mainly based on average differences in IQ. As a result, eugenics was applied to those considered inferior to prevent them from reproducing. Paradoxically, these pioneers of eugenics, advocates of a *sui generis* Social Darwinism,[7] presented themselves as benefactors of humanity who were merely applying practices justified by what they believed to be irrefutable evidence, supported by a form of scientific fundamentalism. Among these pioneers of statistics, inspired by the Galtonian School, who worked in the United States, we should mention

[7] Social Darwinism refers to the somewhat naïve attempt to interpret social phenomena based on Darwinian evolutionary mechanisms (struggle for survival, natural selection) in a very schematic way.

Henry H. Goddard (1866–1957), who worked tirelessly to establish that the majority of European immigrants were feeble-minded,[8] and Lewis Madison Terman (1877–1956), who standardized the Binet test by creating the Stanford-Binet test, suitable for measuring the masses. During *World War I*, he selected millions of American soldiers and, like Goddard, supported the inferiority of European and Mediterranean ethnic groups. He was convinced that the mentally deficient exhibited vicious and violent behavior and, again based on IQ measurements, promoted laws in California that favored sterilization. Finally, Robert M. Yerkes (1876–1956) developed intelligence tests for American soldiers (Army alpha and Army beta) and argued that Mediterranean immigrants should be blocked and selected due to their intellectual inferiority, also promoting restrictive and anti-immigration laws (Carson, 2007). American psychological sciences, therefore, began to promote the blocking of immigration, both to stop hybridization, which was viewed negatively, and to save money for the states by excluding from social and educational services those citizens who, according to the testing method, were considered uneducable (Chase, 1977).[9] One of Yerkes' colleagues, Carl C. Brigham (1890–1943), became infamous for promoting laws in the 1920s that would block immigration in order to halt the decline of American intelligence, as he expressed in a famous book:

> We must now frankly admit the undesirable results which would ensue from a cross between the Nordic in this country with the Alpine Slav, with the degenerated hybrid Mediterranean, or with the negro, or from the promiscuous intermingling of all four types. Granted, the undesirable results of such an intermingling, is there any evidence showing that such a process is going on? Unfortunately the evidence is undeniable [...].
> According to all evidence available, then, American intelligence is declining, and will proceed with an accelerating rate as the racial admixture becomes more and more extensive. The decline of American intelligence will be

[8] Goddard was the best-known psychologist working at Ellis Island in immigrant selection, where Howard A. Knox (1885–1949), a lesser-known pioneer of intelligence testing, also worked. See Zenderland (2001).

[9] As described in the following chapter, during the Fascist period in Italy, an opposing concept spread regarding a kind of positive eugenics called Bonifica umana razionale (Rational human reclamation), which favored intermarriage between northern and southern Italians with the belief that this would result in a genetically improved generation of Italians.

> more rapid than the decline of the intelligence of European national groups, owing to the presence here of the negro. These are the plain, if somewhat ugly, facts that our study shows. The deterioration of American intelligence is not inevitable, however, if public action can be aroused to prevent it [...].
> Immigration should not only be restrictive but also highly selective [...] The really important steps are those looking toward the prevention of the continued propagation of defective strains in the present population. (Brigham, 1923, pp. 208–210)[10]

Meanwhile, in England, Cyril L. Burt (1883–1971), a direct student of Galton, became famous for falsifying the data from his studies on twins in order to support innatist theories, while some of his students, such as Raymond Cattell (1905–1998),[11] Hans Eysenck (1916–1997), and Arthur Jensen (1923–2012) were all renowned academic psychologists, but they too—as we shall see—were involved in controversies concerning their innatist and probably scientistic conception of psychology, going down in history for the ideological positions of their research. Eysenck and Jensen argued that there was a genetically determined general intelligence and that, due to unproven genetic causes, the average IQ of African American populations was significantly lower than that of whites. In the eyes of eugenicist psychologists, this difference justified the existence of a psychological hierarchy between races. Eugenics, psychometrics, and racism thus became a combination whereby the measurement of intelligence was used as a criterion for racializing people. These psychologists were convinced that there were objective intellectual differences between Whites and Blacks, measured by mental tests. Facial angles, craniometric, and anthropometric measurements were thus replaced by psychometrics and psychological tests, which were applied to the entire population using statistical methods.

[10] Carl C. Brigham played a vital role in the history of selection based on intelligence measurements, moving within a few years from an entirely racist and eugenicist approach to intelligence to a highly critical and environmentalist position, eventually stating in the 1930s that "The test movement came to this country some twenty-five or thirty years ago accompanied by one of the most glorious fallacies in the history of science, namely, that the tests measured native intelligence purely and simply without regard to training or schooling. I hope nobody believes that now" (Saretzky, 1982, p. 16).

[11] Cattell was later an important exponent of American psychology, but the link with eugenics always influenced his research work (see Tucker, 2009).

References

Addeo, F. (2008). *È normale la curva normale?* [Is the normal curve normal?]. Bonanno.

Brigham, C. C. (1923). *A study of American intelligence*. Princeton University Press.

Broberg, G., & Roll-Hansen, N. (1996). *Eugenics and the welfare state: Sterilization policy in Denmark, Sweden, Norway, and Finland*. Michigan State University Press.

Carson, J. (2007). *The measure of merit: Talents, intelligence, and inequality in the French and American republics, 1750–1940*. Princeton University Press.

Chase, A. (1977). *The legacy of Malthus: The social costs of the new scientific racism*. Knopf.

Cicciola, E. (2019). *La scoperta dell'intelligenza. Alfred Binet e la storia del primo test* [The discovery of intelligence: Alfred Binet and the story of the first test]. Fefè.

Colla, P. S. (2000). *Per la nazione e per la razza: cittadini ed esclusi nel modello svedese* [For the nation and for the race: Citizens and excluded in the Swedish model]. Carocci.

Danziger, K. (1990). *Constructing the subject: Historical origins of psychological research*. Cambridge University Press.

Darwin, C. (1859). *On the origin of species by means of natural selection or the preservation of favored races in the struggle for life*. Murray.

Foschi, R. (2019). L'eugenetica nella storia della psicologia [Eugenics in the history of psychology]. *Physis. Rivista Internazionale di Storia della Scienza, 1*(2), 291–307.

Foschi, R. (2020). *Storia della psicologia e della mente* [History of psychology and the mind]. Mondadori.

Foschi, R., & Cicciola, E. (2006). Politics and naturalism in the 20th century psychology of Alfred Binet. *History of Psychology, 9*(4), 267–289.

Galton, F. (1883). *Inquiries into human faculty and its development*. Macmillan.

Galton, F. (1889). *Natural inheritance*. Macmillan.

Galton, F. (2001). The Eugenic College of Kantsaywhere. *Utopian Studies, 12*(2), 191–209. (Original work published 1911).

Gillham, N. W. (2001). *A life of Sir Francis Galton: From African exploration to the birth of Eugenics*. Oxford University Press.

Gould, S. J. (1996). *The mismeasure of man* (2nd ed.). Norton.

Hall, L. A. (1990). Illustrations from the Wellcome Institute Library: The Eugenics society archives in the contemporary medical archives centre. *Medical History, 34*(3), 327–333.

Kevles, D. J. (1985). *In the name of eugenics: Genetics and the uses of human heredity*. University of California Press.

La Vergata, A. (2009). *Colpa di Darwin? Razzismo, eugenetica, guerra e altri mali* [Darwin's fault? Racism, eugenics, war, and other evils]. UTET.

Leahey, T. H. (1980). *A history of psychology: Main currents in psychological thought*. Prentice Hall.

Mayr, E. (1982). *The growth of biological thought: Diversity, evolution, and inheritance*. Harvard University Press.

Ray, L. J. (1983). Eugenics, mental deficiency and Fabian socialism between the wars. *Oxford Review of Education, 9*(3), 213–222.

Rose, N. (1985). *The psychological complex: Psychology, politics and society in England, 1869–1939*. Routledge & Kegan Paul.

Saretzky, G. D. (1982). *Carl Campbell Brigham, the native intelligence hypothesis, and the Scholastic Aptitude Test*. Educational Testing Service, Research Publications R-116.

Tucker, W. H. (2009). *The Cattell controversy: Race, science, and ideology*. University of Illinois Press.

Walsh, R. T., Teo, T., & Baydala, A. (2014). *A critical history and philosophy of psychology: Diversity of context, thought, and practice*. Cambridge University Press.

Zenderland, L. (2001). *Measuring minds: Henry Herbert Goddard and the origins of American intelligence testing*. Cambridge University Press.

CHAPTER 6

Life Unworthy of Life

As La Vergata (2009) has highlighted, even before the Darwinian era, the notion that it was permissible to eliminate an inferior race was pretty widespread. In a tragically paternalistic worldview, it seemed natural to believe that eliminating those deemed unfit for society was not wrong. For many racist evolutionists, this was merely an acceleration of a natural process that would ultimately lead to the survival of the fittest and the extinction of the less adaptable. Racists arrogated to themselves the right to anticipate what they considered a natural process, thereby conceiving the physical elimination of so-called unfit races. Eugenics played a central role in this process.

As previously noted, while positive eugenics aimed to select genetic heritage by promoting marriages among individuals considered superior, negative eugenics prohibited reproduction among the "abnormal." The boundary between positive and negative eugenics is relatively thin. Those diagnosed with behavioral problems were often coerced not to reproduce, and coercive sterilization laws were enacted and enforced for decades in major Western countries, including the Scandinavian nations (Broberg & Roll-Hansen, 1996; Colla, 2000).

In this transition that led to the hegemony of negative and eliminationist eugenics over positive eugenics, it is necessary to briefly present the Italian case. In Fascist Italy, particular eugenic programs were implemented, such as the "Rational human reclamation" (Bonifica umana razionale), designed by the Italian physician and characterologist Nicola

R. Foschi, *Racisms*, https://doi.org/10.1007/978-3-032-22962-5_6

Pende (1880–1970). Pende, correlating anthropometric measurements with glandular functioning, developed a biotypology linking morphological and endocrine measures with psychological temperament. He elaborated a complex hormonal typology of personality based on identifying different physical constitutions, which had a significant influence within Italian scientific culture. Specifically, Pende's differential character classification was predicated on the assumption that each observable behavioral characteristic corresponded to a distinct endocrine gland function (hyperthyroid type, hypothyroid type, hyperpituitary type, hypopituitary type, hyperadrenal type, hypoadrenal type, hypergenital type, hypogenital type, hyperthymic type, and hypoparathyroid type); each type corresponded to a morphological-constitutional facial type, either longilinear (sthenic or asthenic) or brevilinear (sthenic or asthenic).[1] These types were easily racialized and identifiable among various Italian populations and subpopulations. Moreover, Pende was an intellectual organically linked to the Fascist regime, applying this knowledge to serve Italian racism.[2] The rational human reclamation was essentially an attempt to create an improved Italian through interbreeding individuals from the North and South in a territory reclaimed from malarial swamps: the Pontine Marshes (in the Lazio region, south of Rome). Pende intended to utilize hygienic and socio-political practices, such as population transfers from the North to the South, to improve the Italian stock by crossbreeding populations. The land reclamation in the lower Lazio region was to be accompanied by the relocation of northeastern populations, who would then settle in the province of Latina (a new city founded during the fascist era). A human reclamation as a consequence of territorial reclamation:

[1] Among character theories developed in Europe between the nineteenth and twentieth centuries, mainly based on constitutional aspects linked to physical qualities and temperament, mention should be made of Sigaud, Mac Auliffe, De Giovanni, Kretschmer, and Sheldon. In contrast, among typologies derived from differential concepts of psychological and psychophysiological aspects, names include Ribot, Malapert, Heymans-Wiersma, and especially Pavlov. In the twentieth century, these characterologies became true psychosomatic typologies, of which Pende's was only one example. See Lombardo and Foschi (2002).

[2] Pende was rector of the University of Bari when it was dedicated to Benito Mussolini, president of the eugenics section of the Italian Research Council (CNR), and founded the *Istituto Centrale di Bonifica Umana* (Central Institute for Human Reclamation), which later became the Sant'Eugenio Hospital in Rome. For Pende's ideological positions during the Fascist period and personal events in that era, see Israel and Nastasi (1998) and Maiocchi (1999).

> The transplantation onto the redeemed swamp lands [...], of strong men, of healthy family units [...], of colonists [...] selected according to anthropotechnical criteria [...] from regions with more fertile populations [...], guaranteed for the three most necessary qualities for the creation of a strong lineage, namely physical robustness, moral health, and proven fertility [...]. (Pende, 1933, p. 241. See also Protasi & Sonnino, 2003)

Pende was thus a regime scientist with a peculiar history; his name appeared among the signatories of the Fascist Racial Manifesto published in 1938 (Cuomo, 2005).[3] However, he claimed his signature was affixed without his consent. After the war ended, he was believed and somewhat exonerated from being one of the signatories. However, as with Evola, the justifications offered for his racism in retrospect do not diminish his responsibility as a racist ideologue. While Evola distanced himself from Nazi racism by claiming his spiritual racism was less severe, without recognizing that it was a prerequisite and provided the cultural context for scientific racism, Pende was the leading figure of a political racism that employed eugenic matrimonial practices, population transfers from North to South, and biological Italian racial reclamation theories aimed at cleansing and forging the fascist man. This was a multifaceted, pseudoscientific racism serving the totalitarian state. His science was subordinated to politics. Individual freedom thus lost its value, and medicine became entirely ancillary to political power, which could endorse official documents like the Racial Manifesto, effectively supplanting the scientist who was entirely subservient.[4]

[3] The manifesto was not original but extremely clear. It expressed pseudoscientific racism in 10 briefly argued points. It was signed by 10 academic scientists and published under the title *Il Fascismo e i problemi della razza* (Fascism and the problems of race) in Il Giornale d'Italia on July 14, 1938. The ten points were: (1) Human races exist [...] (2) There are large and small races [...] (3) The concept of race is purely biological [...] (4) The population of present Italy is majority Aryan in origin, and its civilization is Aryan [...] (5) The contribution of large masses of men in historic times is a legend. (6) A pure "Italian race" now exists [...] (7) It is time for Italians to proclaim themselves openly racists [...] (8) It is necessary to make a clear distinction between Western European Mediterraneans on one side, and Eastern and African peoples on the other [...] (9) Jews do not belong to the Italian race [...] (10) The purely European physical and psychological traits of Italians must not be altered in any way [...].

[4] Giorgio Israel (1945–2015) has devoted many publications to the issue of Pende's adherence to the fascist manifesto of race.

Moreover, positive eugenics prepared the ground for Nazi eliminationist practices. The common basis was hard hereditarianism and innatism applied to politics and ideology. Some eugenicists were openly racist in this regard. For example, psychologist and biologist Charles B. Davenport (1866–1944) promoted negative eugenics in the United States and collaborated with racist German physicians. Davenport was a key figure in organizing the international eugenics movement, which, arising from the work of Galton and other close relatives of Darwin, was gaining traction in Europe. In 1912, Leonard Darwin (1850–1943), Charles's son, organized the First International Eugenics Congress in London. In the United States, Davenport ran an extensive data collection program at the Eugenics Record Office (ERO) located in Cold Spring Harbor, New York. These eugenicists gathered data on families with physical or psychological issues presumed to be genetically determined. The cases studied were highly varied. Classical examples included human genetics cases such as polydactyly, but Davenport also included all forms of mental "weakness" and any typical trait present in a family. Thus, within maritime families, an entirely hypothetical trait relating to a love of the sea was identified; nomadic peoples were defined as such due to their supposed genetic predisposition for travel, and so on (Jackson & Weidman, 2006, p. 114). The ERO's pseudoscientific approach led to the spread of negative and eliminationist eugenics within the American movement. In 1926, American eugenicists founded the American Eugenics Society (AES), articulating a mission described in a "eugenics catechism" to be disseminated at least among insiders.

In this catechism, negative eugenics preceded positive eugenics, constituting its prerequisite. The manifesto's authors planned first to "cleanse" the population of the inferior through sterilization, segregation, and immigration control, and then to improve the population by favoring the reproduction of the fittest and most selected. In this sense, the distinction between negative and positive eugenics, often employed by historians to highlight different forms of eugenics, appears entirely abstract. The two approaches, combined with an ideology consistent with hard hereditarianism and racism, operated collusively within the worldview of these "social scientists," biologists, physicians, and psychologists. In light of these premises, the catechism concluded:

> Q. What is the most precious thing in the world? A. The human germ plasm. Q. How may one's germ plasm become immortal? A. Only by

> perpetuation through children. Q. What is a person's eugenical duty to civilization? A. To see that his own good qualities are passed on to future generations provided they exceed his bad qualities. If he has, on the whole, an excess of dysgenic qualities, they should be eliminated by letting the germ plasm die out with the individual. Q. How many children must a family bear on the average in order to perpetuate the race, that is, just balance itself between life and death? A. The latest figures for the registration area of the United States is between three and four. In other words, three children will not suffice and four will increase the "line." Q. How may my children know all they now are able to about their own pedigree? A. If the parents have filed as complete a family pedigree as possible at the Eugenics Record Office, Cold Spring Harbor, Long Island, the children may have the information at any time by applying there for it. Q. How much does it cost to file a family pedigree? A. Not a cent except the postage. Even the blanks are furnished free of charge. Q. Is the pedigree then open to the public? A. Absolutely no. It is held confidential. Q. What may I do to spread eugenical information? A. Learn, speak and write. The American Eugenics Society will be glad to help by sending you suggestions on reading, lists of speakers, and a leaflet, "Suggested Program for Clubs." Q. What are the Fitter Families Competitions which are conducted by the Eugenics Society? A. These competitions are held at fairs, and will be extended to other appropriate places. Whole families compete for the trophies offered, on the basis of heredity, physique and mentality. The idea is to inculcate in those interested in taking the examinations, and in those who hear about it, the idea that heredity is the important thing, and that proper attention to the laws of heredity and hygiene will be decidedly worth while. Q. How may I bring one of these competitions to my town? A. Write to the American Eugenics Society, which has now standardized the contests and simplified them so that a few doctors and a supervisor can easily handle them. The American Eugenics Society is anxious to extend the competitions to all who are interested. Q. Where may I get information on what to read in eugenics, without having to go through the library, and then not find just what I want? A. Ask the American Eugenics Society for their condensed bibliography on eugenics. (American Eugenics Society, 1926, on Positive Eugenics)

Eugenics thus became a research domain and field of applications regarded as rich with political and administrative potential. It entered directly into the mainstream of science and governance in the United States. Every eugenicist, from their respective disciplinary vantage point, was expected to contribute to the eugenic improvement of humanity. Madison Grant (1865–1937), a New York zoologist and one of the founders of the American Eugenics Society, as well as a probable author of the eugenic catechism, and an admirer of Francis Galton and Georges Vacher de Lapouge, vigorously defended the idea of the superiority of the Northern European race, thereby earning the favor of Adolf Hitler.

Grant was a conservative with a strong ecological sensibility, funding, and actively participating in organizations dedicated to wildlife and nature preservation. He leveraged this conservative ecological stance to promote a negative eugenics ideology in which the Nordic race predominated; through eliminationist practices—including segregation, sterilization, and anti-immigration policies—Grant sought to preserve the Aryans (Anglo-Saxons) from admixture with other "inferior" populations:

> In concluding this revision of the racial foundations upon which the history of Europe has been based, it is scarcely necessary to point out that the actual results of the spectacular conquests and invasions of history have been far less permanent than those of the more insidious victories arising from the crossing of two diverse races, and that in such mixtures the relative prepotency of the various human subspecies in Europe appears to be in inverse ratio to their social value.
>
> The continuity of physical traits and the limitation of the effects of environment to the individual only are now so thoroughly recognized by scientists that it is at most a question of time when the social consequences which result from such crossings will be generally understood by the public at large. As soon as the true bearing and import of the facts are appreciated by lawmakers, a complete change in our political structure will inevitably occur, and our present reliance on the influences of education will be superseded by a readjustment based on racial values. [...]
>
> We Americans must realize that the altruistic ideals which have controlled our social development during the past century, and the maudlin sentimentalism that has made America "an asylum for the oppressed," are sweeping the nation toward a racial abyss. If the Melting Pot is allowed to boil without control, and we continue to follow our national motto and deliberately blind ourselves to all "distinctions of race, creed, or color," the type of native American of Colonial descent will become as extinct as the Athenian of the age of Pericles, and the Viking of the days of Rollo. (Grant, 1916, pp. 227–228)

According to Grant, it was necessary to restrict the reproductive capacity of the disabled, criminals, mentally ill, and members of inferior races:

> As long as such charitable organizations confine themselves to the relief of suffering individuals, no matter how criminal or diseased they may be, no harm is done except to our own generation, and if modern society recognizes a duty to the humblest malefactors or imbeciles, that duty can be harmlessly performed in full, provided they be deprived of the capacity to procreate their defective strain.
> Those who read these pages will feel that there is little hope for humanity, but the remedy has been found, and can be quickly and mercifully applied. A rigid system of selection through the elimination of those who are weak or unfit-in other words, social failures-would solve the whole question in one hundred years, as well as enable us to get rid of the undesirables who crowd our jails, hospitals, and insane asylums. The individual himself can be nourished, educated, and protected by the community during his lifetime, but the state through sterilization must see to it that his line stops with him, or else future generations will be cursed with an ever increasing load of victims of misguided sentimentalism. This is a practical, merciful, and inevitable solution of the whole problem, and can be applied to an ever widening circle of social discards, beginning always with the criminal, the diseased, and the insane, and extending gradually to types which may be called weaklings rather than defectives, and perhaps ultimately to worthless race types. (Grant, 1916, pp. 46–47)

Eugenics and scientific racism, therefore, were not exclusively the prerogative of Nazism or Fascism, as commonly portrayed, but had a preceding and succeeding history that ran parallel or intertwined. Galtonian eugenics, as well as Lombrosian innatism, contained the seeds of the hard racism of the twentieth century. Moreover, eugenic psychologists from influential English and American universities—such as Davenport at Harvard, Terman at Stanford, Yerkes at Harvard and Yale, and Brigham at Princeton—all supported, based on innatism, various eugenic programs, influencing hundreds of students with their theories and legitimizing more blatantly racist positions. Thus, a significant branch of Anglo-American biologists and psychologists, scholars of intelligence, were deeply involved with scientific racism and eugenics, an indelible stain that remains underrepresented in psychology textbooks and insufficiently studied (Richards, 1997). Notably, Grant's writings were even cited by the defense of Nazi hierarchs at the Nuremberg Trials as justification for their criminal conduct. Only in recent years has there been growing awareness in the United States about the legacy of the American eugenics movement.

In 1924, psychologists led by Goddard, demonstrating a presumed intellectual inferiority of immigrants from the Mediterranean and African Americans, helped promote the Immigration Restriction Act, which curtailed immigration. Subsequently, laws permitting forced sterilization proliferated. Indiana was the first state to enact such a law in 1907, targeting criminals, idiots, and mentally ill individuals, followed by similar laws in seventeen other states during the 1920s. Sterilization supplanted the welfare state for those deemed burdensome to public health and social services. The diagnoses leading to sterilization were varied, with eugenicists convinced that sizable proportions of the general population should undergo this fate. Immigrants were estimated to be 40% feeble-minded, whereas the general American population was reckoned at least 10% deserving of sterilization. These were baseless pseudoscientific assertions. The biologist Harry H. Laughlin (1880–1943) spearheaded sterilization campaigns, advocating legislation for the forced sterilization of "feeble-minded, insane, criminalistic, epileptic, inebriate, diseased, blind, deaf, deformed, and dependent"—including "orphans, ne'er-do-wells, tramps, the homeless and paupers" (Singleton, 2014, p. 124). Laughlin received an honorary doctorate in 1936 from the now-Nazi-controlled University of Heidelberg. His association with Nazism led to his academic disgrace in the United States once it became clear that racism and eugenics were the core ideologies of the Hitler regime. Forced sterilization campaigns in various American states resulted in a staggering 60,000 to 90,000 people being sterilized (Jackson & Weidman, 2006, pp. 117–119). Forced sterilization continued in the United States after World War II and is occasionally still applied in forensic settings. Throughout the long history of racism in the United States, eugenics occupied a prominent place, with forced sterilization disproportionately applied to African Americans and people experiencing poverty, who increased social service expenditures. The political and economic use of eugenics colluded with American racism—a white supremacist ideology—pervasive to this day. As noted, even in socially progressive countries like Sweden, programs to sterilize individuals with psychiatric or intellectual disabilities continued officially until 1976. Some Swedish social democratic economists were ardent proponents of sterilization. These eugenics programs were invariably linked to welfare services in a macabre nexus ostensibly aimed at saving public funds and reorganizing health services (Colla, 2000).[5]

In this context, negative eugenics was increasingly radicalized by Nazi Germany and its racist scientists. The progression from sterilization extended to the elimination of individuals deemed undesirable in order to preserve and enhance the Nordic peoples. Foremost among these victims were Jews, who Hitler sought to annihilate, considering them the principal enemy of his race and the cause of Germany's post–World War I misfortunes. Hitler's autobiography (1925–1927/1940) starkly outlines his evolving antisemitism—from cultural disdain to genocidal intent—that fueled his ideology:

> In my eyes the charge against Judaism became a grave one the moment I discovered the scope of Jewish activities in the press, in art, in literature and in the theatre. All unctuous protests were now more or less futile. One needed only to look at the posters and to study the names of the authors of the appalling productions advertised as being performed in the cinemas and theatres, in order to become hardened against the Jews. Here was a pestilence, a moral pestilence, with which the public was being infected. It was worse than the black plague of long ago. And in what doses this poison was manufactured and distributed! [...]
> It was a terrible thought, and yet one which could not be overlooked, that the greater number of the Jews seemed specially destined by Nature to play this shameful part.
> Was it for this reason that they were called the chosen people?
> I then began to investigate carefully the names of all the fabricators of these unclean products which played such a big part in the cultural life of the public. The result of that investigation was still more unfavourable to the attitude which I had hitherto held in regard to the Jews. Though my feelings might rebel a thousand times, reason now had to draw its own conclusions.

[5] It is important to note that from a public finance perspective, nothing was saved; only loss and grievances were created. Genetics is much more complex and also dependent on environment and individual history than eugenicists thought, so it was not possible to "cleanse" humanity via sterilization. Genes may not be expressed in some generations, while unknown mechanisms govern their expression in others. Gene expression and interaction depend on a complex array of factors. These programs failed to achieve their objectives and also caused severe mental disorders in those who had been forcibly sterilized, which a democratic society had to resolve at additional cost. Eugenics was a total failure from a scientific, moral, and financial point of view.

The fact that nine-tenths of all the 'smutty' literature, artistic 'tripe' and theatrical banalities, had to be charged to the account of a people who formed scarcely one per cent of the nation could not be gainsaid. It was there and had to be admitted. (pp. 57–58)
[...]
Making an effort to overcome my natural reluctance, I tried to read articles published in the Marxist press. However, as my aversion increased tenfold, I set about learning something of the people who wrote and published this mischievous stuff — from the publisher downwards, all of them were Jews! (p. 61)
[...]
From time immemorial, however, the Jews have known better than any others how falsehood and calumny can be exploited. Is not their very existence founded on one great lie, namely, that they are a religious community, whereas in reality they are a race? And what a race! (p. 243)
[...]
Hence it is that at the present time, the Jew is the chief agitator for the complete destruction of Germany. Whenever we read of Germany being attacked in any part of the world the Jew is always the instigator. In peace-time as well as during the War the Jewish-Marxist stock-exchange press systematically stirred up hatred against Germany, until one State after another abandoned its neutrality and placed itself at the service of the Allies in the World War, even against the real interests of its own people. (pp. 666–667)
[...]
In this field, too, the National Socialist Movement has a tremendous task before it.
It must open the eyes of our people in regard to foreign nations and it must continually remind them of the real enemy who menaces the world today. Instead of preaching hatred against Aryans from whom we may be separated on almost every other count, but with whom the bond of kindred blood and the main features of a common civilization unite us, we must arouse general indignation against the malevolent enemy of humanity and the real author of all our sufferings.
The National Socialist Movement must see to it that at least in our own country the mortal enemy is recognized and that the fight against him may be the beacon-light of a happier era, and show other nations, too, the way of salvation for struggling Aryan humanity (p. 685).
[...]

> If, at the beginning of the War, or even during the War, twelve or fifteen thousand of these Jewish corruptors of the people[6] had been forced to submit to poison-gas, just as hundreds of thousands of our best German workers from every social class and from every trade and calling had to face it in the field, then the millions of sacrifices made at the front would not have been made in vain. (p. 727)

Following the Jews, Germany's Nazi enemies included disabled individuals, the mentally ill, homosexuals, Roma, political opponents—including Freemasons, who since the eighteenth century had promoted democratic societies and principles of liberty and equality which, however contradictorily interpreted, had germinated abolitionist and labor rights movements, as well as civil rights which the Nazis sought to eradicate along with their proponents.

While the planning of the Holocaust was a radical, entrenched aspect of Hitler's ideology, leading to the systematic elimination of the entire Jewish people through internment and extermination in death camps—resulting in the deaths of at least six million people—the Nazi eugenics programs followed a similarly tragic yet somewhat distinct and less well-known path. The Jewish Holocaust was devised as the "final solution" aimed at the complete eradication of the Jewish population. An operation planned down to the smallest detail, ranging from knowledge of Jewish culture for the purpose of capture, to the rounding up of entire populations in major European cities, to internment, killing, and cremation of bodies as in factory-like death camps (Fig. 6.1) (Hilberg, 1961). This genocide was deemed necessary for the hegemony of the "Aryan" populations. It was implemented alongside anti-Jewish laws passed in Germany (1935) and Italy (1938), which stripped Jews of civil rights primarily through marginalization, expropriation, and dismissal from jobs.

Nazi negative eugenics was directly linked to American eugenics. The goal was to improve the "Aryans" by eliminating "abnormal" elements that would contaminate their genetics. In this regard, their aims were identical to those of American eugenicists.[7]

[6] The reference here is to Marxists, who were the other pole of Hitler's hatred; at the same time, Marxists were identified with Jews by the Nazis.

[7] In Italy, racism was characterized not only by Pende's positive eugenics but also by an obsession with the purity of lineage. Especially when Italian colonialism controlled parts of Africa, segregationist policies were implemented. Cf. Poidimani (2009).

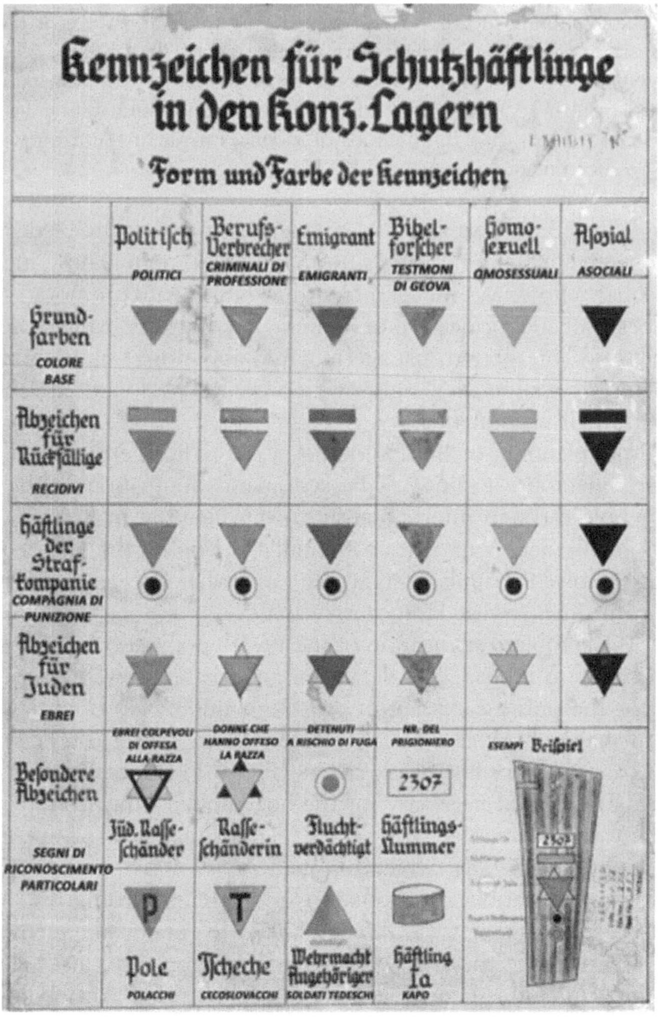

Fig. 6.1 Identification mark for prisoners in Nazi concentration camps (Image: Arolsen Archives, Bad Arolsen)

In Germany, eugenics actually began to spread before Nazism in the early twentieth century, with the foundational idea of Nordicism. The term *Rassenhygiene* (racial hygiene) was coined, and in 1905 Alfred Ploetz (1860–1940) founded the German Society for Racial Hygiene (*Deutsche Gesellschaft für Rassenhygiene*). This society had international contacts and, during the Nazi regime, grew to thousands of members. Its initial purpose was to select and reinforce the "Nordic race" through sterilization, considered superior to Mediterranean, Jewish, Slavic, and nomadic populations. Anglo-American biologists thus inspired German doctors and biologists. In the 1940s, the Nazis more vigorously implemented sterilization and mass extermination programs targeting adults and children diagnosed as dangerous to society due to mental "weakness" or alleged psychopathologies. Many biologists, psychiatrists, and psychometricians lent their expertise and technologies to the innatist cause, while American philanthropists even funded Nazi universities and institutions involved in eugenics and racism; notably, the Rockefeller Foundation supported Nazi eugenic medical institutions during the 1930s.

In 1917, the American propaganda film *The Black Stork* portrayed negative eugenics as the solution for societal well-being (Pernick, 1996). Similarly, the Nazis produced propaganda films (*Erbkrank* [Hereditary Diseased], *Das Erbe* [Heritage], *Opfer der Vergangenheit* [Victims of the Past], *Dasein ohne Leben* [Existence Without Life], *Ich klage an* [I Accuse], etc.) depicting individuals with mental or physical disabilities as non-human, dangerous, ill, or useless to society; science was presented as a disinterested discipline aiding in their elimination for their own good—mirroring Lombroso's proposals for Misdea. These propaganda films promoted the Aktion T4 program, a forced euthanasia campaign using conventional methods such as gas chambers and lethal injections, as well as more insidious methods like progressive starvation in designated institutions. Many Germans placed their trust in the eugenics program, even entrusting it with their children who suffered from physical or psychological disorders, or simply for economic or social reasons.

A chilling feature of this history is that many leading proponents of German racist biologism managed to retain academic positions after WWII and remain active in the scientific community during the Cold War. Among many, some stand out for their role in implementing Aktion T4. Ernst Rüdin (1874–1952), perhaps the most infamous racist psychiatrist, a protégé of Eugen Bleuler (1857–1939) and Emil Kraepelin (1856–1926)—the father of modern descriptive psychiatry—and a relative of

Ploetz, was a key figure advocating psychiatric eliminationism. He became the primary architect of the genocide of the mentally ill, including children, purportedly to improve the Nordic race. After gaining fame and political power, Rüdin authored and oversaw Nazi sterilization laws—such as the "Law on the Prevention of Hereditary Diseases" (Gesetz zur Verhütung erbkranken Nachwuchses)—inspired by American eugenic principles, passed in 1933 and only repealed in 2007 (!). Rüdin also financially supported colleagues who operationalized his ideas through the Aktion T4 program. The name Aktion T4 derives from the administrative address of the program's headquarters. In 1939, Hitler authorized the forced killing of disabled children after receiving a request from a pair of parents to eliminate their son mercifully. This marked the beginning of the rhetoric of "life unworthy of life" (lebensunwertes Leben). The program was applied explicitly in concentration camps as Aktion 14f13 (Lifton, 1986). Until recently, approximately 70,000 deaths from this euthanasia program were estimated, but archival research now reveals figures reaching hundreds of thousands. After the war, only a few of the implicated doctors were prosecuted. Ongoing historical research continues to uncover horrific details, such as the involvement of Hans Asperger (1906–1980), a Viennese pediatrician previously believed to oppose Nazism and the Aktion T4 program. Asperger had developed the concept of high-functioning autism (Asperger syndrome) precisely to broaden autism diagnosis thresholds to accommodate the Nazi eugenics agenda, which sought to include as many young patients as possible—a move that aided his career (Czech, 2018; Sheffer, 2018). These revelations have sparked scientific debate about whether the term Asperger syndrome should continue in use (Baron-Cohen, 2018).

References

American Eugenics Society. (1926). *A eugenics catechism*. American Eugenics Society.

Baron-Cohen, S. (2018). Hans Asperger's Nazi collusion. *Nature, 557*, 305–306.

Broberg, G., & Roll-Hansen, N. (1996). *Eugenics and the welfare state: Sterilization policy in Denmark, Sweden, Norway, and Finland*. Michigan State University Press.

Colla, P. S. (2000). *Per la nazione e per la razza: cittadini ed esclusi nel modello svedese* [For the nation and for the race: Citizens and excluded in the Swedish model]. Carocci.

Cuomo, F. (2005). *I dieci. Chi erano gli scienziati italiani che firmarono il "Manifesto della razza"* [The ten: Who were the Italian scientists who signed the "Race Manifesto"]. Baldini Castoldi Dalai.

Czech, H. (2018). Hans Asperger, national socialism, and "race hygiene" in Nazi-era Vienna. *Molecular Autism, 9*, 1–43.

Grant, M. (1916). *The passing of the great race; or, the racial basis of European history*. Charles Scribner's Sons.

Hilberg, R. (1961). *The destruction of the European Jews*. Quadrangle Books.

Hitler, A. (1940). *My Struggle*. Stalag Edition. (Original work published 1925–1927)

Israel, G., & Nastasi, P. (1998). *Scienza e razza nell'Italia fascista* [Science and race in Fascist Italy]. Il Mulino.

Jackson, J. P., & Weidman, N. M. (2006). *Race, racism, and science: Social impact and interaction*. Rutgers University Press.

La Vergata, A. (2009). *Colpa di Darwin? Razzismo, eugenetica, guerra e altri mali* [Darwin's fault? Racism, eugenics, war, and other evils]. UTET.

Lifton, R. J. (1986). *The Nazi doctors: Medical killing and the psychology of genocide*. Basic Books.

Lombardo, G. P., & Foschi, R. (2002). *La costruzione scientifica della personalità* [The scientific construction of personality]. Bollati Boringhieri.

Maiocchi, R. (1999). *Scienza italiana e razzismo fascista* [Italian science and fascist racism]. La Nuova Italia.

Pende, N. (1933). *Bonifica umana razionale e biologia politica* [Rational human reclamation and political biology]. Cappelli.

Pernick, M. S. (1996). *The Black Stork*. Oxford University Press.

Poidimani, N. (2009). *Difendere la razza. Identità razziale e politiche sessuali nel progetto imperiale di Mussolini* [Defending the race: Racial identity and sexual policies in Mussolini's imperial project]. Sensibili alle foglie.

Protasi, M. R., & Sonnino, E. (2003). Politiche di popolamento: colonizzazione interna e colonizzazione demografica nell'Italia liberale e fascista [Settlement policies: Internal colonization and demographic colonization in liberal and fascist Italy]. *Popolazione e storia, 4*(1), 91–138.

Richards, G. (1997). *'Race', racism and psychology: Towards a reflexive history*. Routledge.

Sheffer, E. (2018). *Asperger's children: The origins of autism in Nazi Vienna*. Norton.

Singleton, M. M. (2014). The 'Science' of eugenics: America's moral detour. *Journal of American Physicians and Surgeons, 19*(4), 122–126.

CHAPTER 7

Discriminations

The title of this volume is plural because, as we have seen, racism—including its strictly biological forms—has manifested in various ways, all shaped by the logic of hierarchy among human types. Eugenics served as a means to implement policies inspired by racist ideologies aimed at controlling the "inferior." Racism was grounded in cultures of discrimination, and the underlying logics of these cultures are consistent across all historically racialized groups. Some historians prefer to keep racism separate from other forms of discrimination. This tendency leads to a superficial treatment of the phenomena, overlooking the significant analogies between racism and discrimination. Simplifying racism flattens its complexity; instead, racism proves to be pervasive, complex, and multifaceted.

Colette Guillaumin (1934–2017) was among the first to focus on the interactions between sexism and racism, highlighting the similarities between the two forms of discrimination, which she also considered to be a true, pervasive ideology, typical of all discrimination (Guillaumin, 1995). Following this line of critical thinking, jurist Kimberle Crenshaw introduced the concept of "intersectionality," whereby each individual can be racialized multiple times as belonging to multiple racialized groups. For example, the same individual can be Jewish, female, Lesbian, Black, and so on. The subject, therefore, suffers a sort of multiple marginalization, including multiple forms of legal discrimination, because

R. Foschi, *Racisms*, https://doi.org/10.1007/978-3-032-22962-5_7

they are classified in more than one racialized category (Crenshaw, 1989; see Ciracì, 2024).

Gender was among the first categories subjected to discrimination according to logics very similar to those characteristic of racism. Women were considered biologically inferior to men, just as African Americans were deemed inferior to Northern Europeans. In this sense, although the concepts of race and gender differ, racism and sexism operate through similar mechanisms. Discrimination is the key concept that follows analogous logics across all types of categorical distinctions. Today, beyond sexism, discussions encompass classism (discrimination based on social class), homotransphobia (discrimination and aversion toward homosexuality and transgender identities), ableism (discrimination against disabilities), ageism (discrimination based on age, especially targeting aged people), heterosexism (preference for heterosexuality), misoneism (hostility to innovations), and various cultural, religious, and youth-related discriminations, among others.

All these longstanding forms of discrimination are based on traditional prejudices often founded on empirical appearances. Regarding notions of female inferiority, Darwin himself is once again invoked for having expressed sexist ideas framed as scientific reasoning. Observing that almost all scientists, artists, and writers were male, Darwin concluded that evolution had produced gender superiority. *In The Descent of Man* (1871), Darwin argued that man's evolutionary role in conquering the female led to male predominance (Saini, 2017). Craniology later advanced the idea that women were inferior to men because they had smaller brains and skulls in terms of mass, weight, or circumference. Two particularly telling excerpts from Broca and one of his students, Paul Topinard (1830–1911), cited by Stephen Jay Gould in *The Mismeasure of Man* (1996), state:

> We might ask if the small size of the female brain depends exclusively upon the small size of her body. Tiedemann has proposed this explanation. But we must not forget that women are, on the average, a little less intelligent than men, a difference which we should not exaggerate but which is, nonetheless, real. We are therefore permitted to suppose that the relatively small size of the female brain depends in part upon her physical inferiority and in part upon her intellectual inferiority. (Broca, 1861, p. 153)
> The man who fights for two or more in the struggle for existence, who has all the responsibility and the cares of tomorrow, who is constantly active in

> combatting the environment and human rivals, needs more brain than the woman whom he must protect and nourish, than the sedentary woman, lacking any interior occupations, whose role is to raise children, love, and be passive. (Topinard, 1888, p. 22)

The universally held, apodictic idea was that women possessed a precise biological role, rendering them inferior to men both intellectually and physically. Since intelligence was believed to correlate with brain mass, smaller brain weight and circumference were empirically interpreted as evidence of female inferiority. Women were also viewed as naturally passive and submissive because nature assigned them a specific, predetermined fate:

> The male human was given the aggressive mission; the female human the problematic task of defense. The part assigned to man is simple and requires just strength—be it physical, moral, intellectual, or the complex combination of many elements—but always energy for attack and seduction: to assault, attack, overthrow, one after another, the curtains and bulwarks, barricades and lunettes, all the complicated apparatus of fortification that the woman will raise against him to defend herself or, better said, to let herself be conquered slowly and modestly.
> The woman, however, had a much more cruel and challenging task assigned by nature. She must deny what she desires, must struggle against the lust that invades her; must reject the one she loves; must demand sacrifices but would rather kiss; must be stingy when everything urges her to generosity; must gather all her meager strength to defend a fiercely assaulted gate, and must loudly shout "wait!" to the one she would tenderly hold to her breast. (Mantegazza, 1873, pp. 82–83)

Just as enslaved African Americans were denied equal dignity and rights because of presumed moral and intellectual inferiority, women were similarly denied equal rights because they were considered biologically and intellectually inferior (Babini, 1999). Various aspects of female corporeality were studied expressly to justify the subordinate role ascribed to women, even in a modernizing society that was embracing scientific ideals (Babini et al., 1986). Despite the emergence of new ideas based on the Enlightenment and the concept of progress, women were relegated to a subordinate role. As an example in the Italian legislative debate regarding the electoral reform between 1880 and 1881, which proposed franchise expansion, Giuseppe Zanardelli (1826–1903)—a progressive politician in many respects[1]—declared:

> Woman is different from man; she is not called to the same offices, nor to militant public life; her place is the family, her life is domestic, her characteristics are the affections of the heart that do not reconcile with the duties of civil life; her mission is to form patriotic, liberal citizens; and *domi mansit lanam fecit* (she looked after the house, she spun wool) must be the most honorable motto of her program; the strength of woman is not in public meetings, but in the empire of heart and feeling over cold calculation and cruel reason. (Zanardelli, 1881, p. 5035)

The positions on women held by positivists reflected ideas that, in a cruder form, were widespread throughout Western society. Lombroso gave a pseudoscientific veneer to discrimination against women, and, together with his son-in-law, Guglielmo Ferrero (1871–1942), authored a treatise on women that indulged in pseudonaturalistic observations justifying what seemed to most positivists to be self-evident truths about female inferiority and their role, ones that science purportedly demonstrated with ease.

The treatise by Lombroso and Ferrero, which aims to describe female criminal behavior, devoted a substantial first part to the "normal" woman in order to compare men, women, and then the differences between female and male criminals. Thus, they described a range of animals in which females exhibited superior biological traits. However, upon reaching mammals, the authors decisively stated that "the power of the male is without contest" (Lombroso & Ferrero, 1893, p. 8). For Lombroso, as for other evolutionists, the perceived superiority was due to men developing and selecting strength as a requirement for survival, which elevated them as organisms above women, who were not required to perform such activities:

> However, the struggle among males—struggle originating from another fact, the greater ardor of sexual desires and perhaps also from the greater

[1] Zanardelli, a prominent figure in Italian Freemasonry, was the first Prime Minister to promote systematic funding for the development of southern Italy. In the second half of the nineteenth century, he had already drafted a liberal penal code that abolished the death penalty and later presented bills on sex education and divorce. He was a supporter of the right to strike in Italy.

> number of individuals—at higher orders developed their size and strength to make them bigger and stronger and increased their primitive superiority in structure, by the effect of organ use and selection, combined with the biological law, highlighted by Spencer, of antagonism between reproduction, growth, and structure.
> Throughout the series, the male has, in short, a potential for development superior to the female, due to the lesser importance of his participation in reproduction [...]. It is therefore understood how, under the influence of life conditions, the male, previously smaller, has, by a biological law, been capable of developing beyond the female. The male, therefore, is a more perfect and variable form of the female due to the greater development of secondary sexual characteristics. (p. 13)

The treatise continued to highlight physical measurements starting with the skull and brain, presenting abundant data showing women to be inferior to men in all measured respects. A chapter on female sensitivity aimed to demonstrate a lower activation threshold, especially regarding sexuality, which for the "normal" woman involved little sexual interest and difficulty achieving orgasm:

> Being naturally and organically monogamous and frigid, it is understandable why laws against adultery have affected women in almost all societies, and not men, who have too often had to circumvent them; and this explains, if not justifies, the eternal injustice with which the law and custom treat women compared to men in marital relationships.
> It is needless to recall that what is not even a misdemeanor for the male is a very grave crime for the woman.
> It also explains how prostitution, which should be legally and socially tolerated for many reasons (relief of male ardor, prevention of crime), has gradually become a cause for infamy among civilized peoples, and how it exists, one might say, regularly, officially only for the benefit of men. In contrast, it hardly exists for women, who, from a biological perspective, would not need it. (pp. 57–58)

Prostitution was thus considered biologically necessary for men, and unnatural for women. In this pseudoscientific ideology, women evolved solely as mothers without possessing a sexuality requiring expression. The authors wrote:

> Therefore, organically, the woman is a mother more than a lover of a man. Although sometimes she is very attached and devoted to her husband, it

> is for other indirect reasons, more than for sexual sympathy. The woman seeks the man, besides the maternal need, for the need of support. (p. 126)

Given women's limited sensitivity, the authors further argued that male beauty was less important to women, who, in seeking support, would prefer strength over beauty.

Regarding moral qualities, Lombroso and Ferrero stubbornly attempted to demonstrate, based on tautological and pseudoevolutionist reasoning, that women tended toward lying, vanity, lack of conscientiousness, anger, greed, disloyalty, dishonor, envy, misoneism (conservatism and hostility to progress), low creativity, and even intellectual inferiority.

The lower intelligence of women was always explained by men's need to struggle for survival, which purportedly led them to devise strong and creative behaviors, whereas women led a dull, passive life, and motherhood was seen as a conservative and undynamic occupation. In Lombroso, the "normal" woman was thus wholly inferior to the man and placed on the same level as the savage, the criminal, and the child:

> The average woman has many traits that bring her closer to the savage, to the child, and therefore to the criminal (irritability, revenge, jealousy, vanity), and other diametrically different positive traits that neutralize the negative ones but prevent her from approaching the balance between rights and duties, selfishness and altruism, which is the goal of moral evolution and characteristic of man. (p. 157)

As we have seen, discussing scientific racism, crowd psychology theorists such as Le Bon also viewed the crowd as regressed and in need of taming, likening it to a woman. It is striking that contemporary historians have devoted whole books to the second part of Lombroso's treatise, specifically on the female criminal who was mainly identified as the prostitute (Azara & Tedesco, 2020; Montaldo, 2019), while neglecting the first part of the book dedicated to the "normal" woman, who, as we have seen, was essentially inscribed as a sort of inferior race relative to the Western white man.

The "normal" woman was described as full of anomalies and inferiority, differing from the female criminal only in degree of excess; female criminality was ultimately taken for granted for a gender regarded as so imperfect. The woman was considered in a separate and inferior "race"; female criminals and prostitutes, compared to normal women, exhibited only more exuberant sexuality and passion, which supposedly led them to delinquency.

Prejudice thus formed a heavy fetter on women's lives, and those who attempted emancipation were met with sarcasm and disdain on all levels. Guido Della Valle (1884–1962), a Neapolitan academic and pedagogue, reviewed Maria Montessori's pedagogical work at the Casa dei Bambini as follows:

> Believe me, dear lady: it is funnier to relate simply the chatters with the little tenants of the philanthropic Roman building institute rather than reading those cold German or English books, so packed with experimental figures and serious questions concerning those problems that you consider so simple. Do not envy us, poor Philosophical Doctors, our melancholy. On the contrary, as a good Medical Doctor, please go back to anthropology, pediatrics, legal medicine, obstetrics, hygiene, ... to the kitchen [italics in text]. (Della Valle, 1911, p. 80. See also Foschi, 2008)

If women's racialization was a logical consequence following recognition of sex differences with distinct phenotypes, sexuality sciences began constructing typifications when positivism took root, and around 1870, the medical-scientific category of homosexuality emerged. For homosexuals, as for women, nineteenth-century scientists expressed pseudoscientific considerations that were similar to the attributions that identified inferior races. To demonstrate the inferiority of inferior races, colonizing "scientists" pointed to homosexuality or hypersexuality as evidence (De Leo, 2021). By analogy, Western homosexuals were considered similar to individuals belonging to inferior races and therefore susceptible to being racialized.

According to Hacking (2005), since the 1870s, homosexuality has been treated with different aims, both essentialist and constructionist. Essentialists emphasized genetics in determining the biological existence of homosexuals; constructionists believed biological data research was futile and yielded contradictory results, deeming history and context more important to highlight distinct gender identities.

During the nineteenth century, the essentialist and genetic approach predominantly promoted prejudice and typification. While currently essentialism and the search for a homosexuality gene may be used to affirm that homosexuals "are who they are" without cultural determinism, in the nineteenth century, essentialism and search for the homosexuality gene were viewed as a defect designating abnormality, inferiority, and irresponsibility. Homosexuality transitioned from being considered an immoral act, a sin to be expiated, to being deemed in positivist times an imperfect and pathological sexuality compared to heterosexuality:

> At that time, homosexual acts between men, such as sodomy, were punished by increasingly strict enforcement of the law. Homosexuality was considered a vice, and homosexual practices were condemned, causing many homosexuals to feel guilty. According to Krafft-Ebing, on the contrary, homosexuals were not immoral or evil men, but men who suffered from a perversion of normal instincts. Homosexuals were perverts, and perverts were sick. This classification may seem oppressive to us and sound like a condemnation, but at the time, doctors considered it a liberation from moral stigma. It was not until around 1980, thanks to the political action of homosexuals themselves, that psychiatrists reconsidered this classification; they finally admitted that homosexuality is not a disease. (Hacking, 2005, Un cadre d'analyse [An analysis framework], p. 9)[2]

While women were considered inferior, homosexuals were viewed as degenerates and abnormalities comparable to criminals. Following Lombroso's thought helps us understand the roots of this pseudoscientific prejudice. Referring to contemporaneous psychopathologists who treated homosexuality as a disease, Lombroso wrote an article comparing innate criminality and homosexuality:

> However, the psychology of both is always the same, amoral and often criminal, and above all, strange.
> Among homosexuals, one notes frivolity, selfishness, jealousy, falseness, lying, gossip, vanity for outward adornments, a particular aesthetic inclination, and, for this, one sees among them a significant number of theater people and also great artists, painters, and musicians.
> Similarly, among criminals, there is a propensity for orgy, revenge, zoophilia, love of evil for evil, the use of tattoo jargon, hieroglyphs, all recalling the most extreme atavism.
> Both homosexuals and born criminals have an analogous, if not identical, etiology.

[2] As will be seen in the next chapter, homosexuality was actually removed from the main psychiatric diagnosis manual (DSM-II) with the seventh printing in 1974.

> Both descend from people with epilepsy, neurotics, neuropathics, eccentric or old parents, although born criminals have a higher number of alcoholic parents.
> The age at which crime causes the most devastation is the same for both, from 15 to 25 years.
> Both criminals and homosexuals have a neuropathic and morbid core, although hysteria is more frequent among the latter, and epilepsy among the former; however, excessive impulsivity, precocity, simulation of madness, and inability to inhibit give both essentially epileptic coloring.
> Both in criminal propensity and inversion, this can be equivalent to an epileptic seizure and seems almost like a cyclical return.
> It is indeed strange that a considerable number of scholars, such as Näcke, Lacassagne, Gross, and Krafft-Ebing, reject as inconceivable the existence of born criminals who have a far greater number of traits. However, all accept the innatism and consequent non-responsibility of homosexuals, who, however, have less evident traits and whose anomalies and misdeeds are fewer.
> It is certain that although one must accept the complete parallelism of born criminals and born homosexuals, both completely incorrigible, social and legal measures against the latter cannot be as severe as against the former, because their misdeeds are far fewer. It can be sure they will cease with loss of sexual activity, which cannot be said of the dangerous criminal until the last day of his life, against whom preventive measures will never be enough. (Lombroso 1906/1985, p. 72)

For Lombroso, Lesbians, like homosexuals, were undoubtedly degenerates to be treated as a subcategory of female criminals. Lombroso amplified the medical and psychiatric positions of the nineteenth century regarding these particular gender identities; for Lesbians, the keywords were degeneration and criminality:

> The influence of degeneration increasingly tends to bring closer and confuse the two sexes, so that among criminals one finds female infantilism in the male leading to pederasty, corresponding to masculinity in women, due to a tendency to revert to the hermaphroditism period.
> The proof is that in many (Lesbians) this tendency preceded even puberty; many enjoyed dressing as men […], loved to see female organs, and shunned women's work. In fact, according to Schüle, frequent cases of sexual perversion are found in moral dementia and epilepsy. (Lombroso & Ferrero, 1893, p. 416; see Fig. 7.1)

Fig. 7.1 Lesbian couple (Lombroso & Ferrero, 1893, p. 423)

During fascism, Lombroso's ideas about criminality were essentially institutionalized. Homosexuality was first treated as a disease and later as a threat to the Italian race. The man fascism aimed to build was virile and the opposite of what homosexuals represented; this posed a political as well as scientific problem. In the 1930s, Italy intensified penalties and exile for homosexuality in line with Nazism (Dall'Orto, 2000).

The Nazis, for their part, were much more explicit in considering homosexuals a category to be eliminated alongside inferior races, whose survival would threaten the health of the youth and the strength of the Nordic-Aryan race. Historians have also shown that, starting in the 1920s, many activists involved in promoting a new tolerance toward sexuality—women's liberation, greater tolerance toward nudism, pornography, and homosexual behavior—were Jewish. Hence, the fascists and Nazis easily attributed all the emerging cultural changes in Western society to the Jews (Herzog, 2011). From the mid-1930s, Nazi laws harshened penalties against homosexuality, explicitly linking it to racial purity needs, and created the *Central Reich Office for the Fight against Homosexuality and Abortion* (Reichszentrale zur Bekämpfung der Homosexualität und der Abtreibung). The victims numbered in the hundreds of thousands, and in concentration camps, homosexuals wore a pink triangle on their uniforms. Moreover, the persecution of homosexuals was supported by the traditionalist branch of the Catholic Church, which already stigmatized homosexuality as unnatural and sinful, and which in Germany endorsed Nazi positions.[3]

After the liberation from Nazifascism and during the Cold War, there emerged another phase of persecution of homosexuality, more directly linked to preexisting homophobic cultures also present in liberal states. Homosexuals continued for a long time to be discriminated against and marginalized, even in cultures that defended civil rights for other categories, such as workers. Psychiatry especially continued to consider homosexuality a pathology until the 1980s.

References

Azara, L., & Tedesco, L. (Eds.). (2020). *La donna delinquente e la prostituta: L'eredità di Lombroso nella cultura e nella società italiane* [The female delinquent and the prostitute: Lombroso's legacy in Italian culture and society]. Viella Libreria Editrice.

Babini, V. P. (1999). Un altro genere. La costruzione scientifica della «natura femminile» [Another gender: The scientific construction of "female nature"]. In A. Burgio (Ed.), *Nel nome della razza. Il razzismo nella storia d'Italia, 1870-1945* [In the name of race: Racism in Italian history, 1870–1945] (pp. 475–489). Il Mulino.

Babini, V. P., Minuz, F., & Tagliavini, A. (1986). *La donna nelle scienze dell'uomo* [Woman in the human sciences]. FrancoAngeli.

Broca, P. (1861). Sur le volume et la forme du cerveau suivant les individus et suivant les races [On the volume and shape of the brain according to individuals and races]. *Bulletin de la Société d'Anthropologie* (Paris), *2*, 139–204, 301–321, 441–446.

Ciracì, F. (2024). *Sul razzismo. Strutture logiche e paradigmi storico-filosofici* [On racism. Logical structures and historical-philosophical paradigms]. Mimesis.

[3] Early right-wing homosexual movements initially believed Nazism might help, as Storm Division (Sturmabteilung, SA) leader Ernst Röhm (1887–1934) was openly homosexual and a personal friend of Hitler from early on. Later, the SA fell from favor, Röhm was forced to commit suicide, and the regime became definitively homophobic, targeting homosexuals and lesbians for elimination like Jews, other "inferior" ethnicities, and political opponents. See Herzog (2011).

Crenshaw, K. (1989). Demarginalizing the intersection of race and sex: A Black feminist critique of antidiscrimination doctrine, feminist theory and antiracist politics. In *University of Chicago Legal Forum, 140*(1), 139–167. Available at: http://chicagounbound.uchicago.edu/uclf/vol1989/iss1/8

Dall'Orto, G. (2000). Il paradosso del razzismo fascista verso l'omosessualità [The paradox of fascist racism towards homosexuality]. In A. Burgio (Ed.), *Nel nome della razza. Il razzismo nella storia d'Italia, 1870–1945* [In the name of race: Racism in Italian history, 1870–1945] (pp. 515–528). Il Mulino.

Darwin, C. (1871). *The descent of man and selection in relation to sex* (2 vols.). Murray.

De Leo, M. (2021). *Queer. Storia culturale della comunità LGBT+* [Queer. A cultural history of the LGBT+ community]. Einaudi.

Della Valle, G. (1911). Le Case dei Bambini e la Pedagogia scientifica di M. Montessori [The Children's Houses and the scientific pedagogy of M. Montessori]. *Rivista pedagogica, 4*, 67–80.

Foschi, R. (2008). Science and culture around the Montessori's first "Children Houses" in Rome (1907–1915). *Journal of the History of the Behavioral Sciences, 44*(3), 238–257.

Gould, S. J. (1996). *The mismeasure of man* (2nd ed.). Norton.

Guillaumin, C. (1995). *Racism, sexism, power and ideology*. Routledge.

Hacking, I. (2005). *Façonner les gens (II). Cours au Collège de France (2004–2005)* [Shaping People (II). Lectures at the Collège de France (2004-2005)]. https://www.college-de-france.fr/fr/agenda/cours/faconner-les-gens-ii

Herzog, D. (2011). *Sexuality in Europe: A twentieth-century history*. Cambridge University Press.

Lombroso, C. (1985). Del parallelismo tra l'omosessualità e la criminalità innata [On the parallelism between homosexuality and innate criminality]. *Sodoma, II*(2), 70–72. (Original work published 1906).

Lombroso, C., & Ferrero, G. (1893). *La donna delinquente, la prostituta e la donna normale* [The female delinquent, the prostitute and the normal woman]. Roux.

Mantegazza, P. (1873). *Fisiologia dell'amore* [Physiology of love]. Bernardoni.

Montaldo, S. (2019). *Donne delinquenti. Il genere e la nascita della criminologia* [Female delinquents: Gender and the birth of criminology]. Carocci.

Saini, A. (2017). *Inferior: How science got women wrong—and the new research that's rewriting the story*. Beacon Press.

Topinard, P. (1888). Le poids de l'encéphale d'après les registres de Paul Broca [Le poids de l'encéphale d'après les registres de Paul Broca]. *Mémoires Société d'Anthropologie Paris,2nd series,* 3, 1–41.

Zanardelli, G. (1881). *Discussioni del 4 aprile 1881* [Discussions of April 4, 1881]. In *Atti del Parlamento italiano - Discussioni della Camera dei Deputati, XIV Legislatura - Sessione 1880-1881 (24/02/1880-09/04/1881), Volume V dal 24/02/1880 al 09/04/1881* [Proceedings of the Italian Parliament—Debates of the Chamber of Deputies, XIV Legislature—Session 1880–1881 (February 24, 1880–April 9, 1881), Volume V from February 24, 1880 to April 9, 1881] (pp. 5001–5016). Tipografia Eredi Botta.

CHAPTER 8

Pride and Reaction

In this chapter, we will examine the history of racism from the perspective of racialized groups, focusing in particular on how these groups have attempted to emancipate themselves. However, we will also discuss conservative reactions against these attempts.

The feeling of Pride, together with the desire for justice, was the driving force of emancipation. Pride concerns the perception of one's own value, which is necessary for affirming a demand for equality on the part of all those racialized and discriminated categories. The labor movement, the Civil Rights movements of African Americans, the women's emancipation movements, the organizations of opposition to antisemitism, and the LGBTQ+ movement would never have been born had they not been based on a proud sense of vindication in favor of their marginalized and persecuted condition.

Prejudices against the subproletariat and the proletariat were similar to those against so-called savage peoples. The clergy, the aristocracy, and certain bourgeois elements feared the "fourth estate," just as they feared "primitive" peoples. The history of workers' rights is a history that begins in the nineteenth century, developing out of other movements founded on Enlightenment ideas regarding justice and equality, which had already brought about the bourgeois revolutions. This period concluded with the founding in 1864 of the *International Workingmen's Association (IWA)*, also known as the *First International*, which was rooted in anarchist and socialist ideals. From that moment on, the

R. Foschi, *Racisms*, https://doi.org/10.1007/978-3-032-22962-5_8

organized labor movement in the West operated in relatively modern and proletarianized societies through militancy, campaigns demanding rights, and revolutionary attempts such as the Paris Commune (1871). As we have seen, however, within the European workers' movement, the racial question was by no means central and, for example, anti-Jewish prejudice, at least before the Dreyfus Affair, was very much present even among the socialists.

In the United States, however, the problem of slavery was more pressing. Slavery was widely practiced; indeed, precisely during the Enlightenment, enslaved Africans were sold to Westerners, and the issue became highly relevant, especially for the progressive ruling class. At the beginning, racism and slavery were even justified by the founding fathers of the United States, such as Thomas Jefferson (1743–1826), who himself possessed hundreds of enslaved people. Jefferson, moreover, was highly contradictory on the issue of equality and slavery. This contradiction derived, among other things, from polygenism, or the view that humans have different origins. Today we know this view to be mistaken, but in those days it was generally accepted among racists to justify their worldview (Jackson & Weidman, 2006). Polygenism seemed rational and made the existence of human races derived from different ancestors more directly plausible. The Bible contains the story of the Curse of Ham, who was considered the precursor of Africans. The other sons of Noah had been more fortunate, and the other, non-cursed races were thought to have descended from them; everything had to remain coherent with these beliefs (see Chapter 1).

Racists were by no means absent among the monogenists either, precisely because, whatever the scientific framework employed, the supposed superiority of some races over others seemed more than obvious. Monogenist racists were often religious and creationist. The Bible itself was easily interpreted as a narrative describing only the creation of white men in the image and likeness of God, according to a monogenist line.

As we emphasized at the opening of this volume, egalitarianism was also one of the values of the Enlightenment, and Jefferson represented one of its foremost North American exponents. In the *Declaration of*

Independence of 1776, he had written the famous phrase "all men are created equal."[1] A fundamental problem indeed.

In his contradictions, Jefferson was therefore one of the figures who gradually opened the way to legislation increasingly attentive to civil rights. In fact, though racist, Jefferson eventually decoupled racism from slavery and began to promote legislation that regulated and mitigated the deportation of enslaved people, while remaining convinced of racial differences. It was a paternalistic policy, as the hierarchical organization of races appeared to be a real phenomenon. Thus, it was only necessary to moderate through legislation what was believed to be a natural fact: the substantial differences among human groups based on race. Perhaps society could also rid itself of slavery, but it was nonetheless necessary to maintain the separation of the races based on pseudoscientific ideas. Segregationism also emerged in this way. It was another problem tied to racism, and appeared in the United States at the end of slavery:

> I advance it therefore as a suspicion only, that the Blacks, whether originally a distinct race, or made distinct by time and circumstances, are inferior to the Whites in the endowments both of body and mind. It is not against experience to suppose, that different species of the same genus, or varieties of the same species, may possess different qualifications. Will not a lover of natural history then, one who views the gradations in all the races of animals with the eye of philosophy, excuse an effort to keep those in the department of man as distinct as nature has formed them? This unfortunate difference of color, and perhaps of faculty, is a powerful obstacle to the emancipation of these people […]. Among the Romans emancipation required but one effort. The slave, when made free, might mix with, without staining the blood of his master. But with us a second is necessary, unknown to history. When freed, he is to be removed beyond the reach of mixture. (Jefferson, 1787, pp. 139–240)

[1] The idea of equality among human beings, as a natural right to be included in the founding documents of the United States, was directly derived from Enlightenment-Masonic thought and likely reached Jefferson via his Tuscan friend Filippo Mazzei (1730–1816), a physician, politician, and Freemason who participated in the American Revolution. Already a friend of Benjamin Franklin (1706–1790), Mazzei moved near Jefferson's estate in the years before the 1776 "Declaration." An essayist and polemicist, he was persecuted in Italy by the Inquisition. In America, his ideas found fertile ground; Jefferson translated some of his writings, whose excerpts appear in American constitutional documents.

The debate around Jefferson and his ambiguity concerning egalitarian values—which remained valid only for Whites—is typical of all those ideas that are welcomed as fundamental by the most open politicians and intellectuals but are not universally applied due to the fear, on the part of the democratic ruling class itself, that accelerating progressivism could prove counterproductive for maintaining power and/or dangerous for society. To liberate but segregate was nonetheless the step that began to disconnect racism from slavery.

The contradictions of these racist Enlightenment figures were thus many, and to understand the contrast among the different positions, it is necessary to mention both Benjamin Rush (1746–1813) and Benjamin Franklin. Rush, a chemist and physician, was another of the founding fathers of the United States who, unlike Jefferson, held positions not only as an abolitionist but also as an anti-racist. For him, there were no moral or psychological differences between Whites and Blacks; nevertheless, he too purchased at least one enslaved person. Franklin, for his part, was one of the leading exponents of eighteenth-century Freemasonry—a man of science and a prominent political figure—who gradually shifted from being a slaveholder (he too owned slaves) to an anti-racist, convinced that slavery should be abolished. In this transformation, he was influenced by acquaintances with important abolitionists active in the *Religious Society of Friends* (Quakers). This religious organization advocated radical egalitarian positions derived from Protestant Christianity, inspired by the earliest Christian communities (Nash, 2006). Above all, the observation that the education of African Americans could render Whites and Blacks morally equal firmly pushed Franklin toward anti-racist positions. This passage from a letter written to a friend is particularly significant in showing the change in Franklin's convictions:

> This is chiefly to acquaint you, that I have visited the Negro School here in Company with the Revd. Mr. Sturgeon and some others; and had the Children thoroughly examin'd. They appear'd all to have made considerable Progress in Reading for the Time they had respectively been in the School, and most of them answer'd readily and well the Questions of the Catechism; they behav'd very orderly, showd a proper Respect and ready Obedience to the Mistress, and seem'd very attentive to, and a good deal affected by, a serious Exhortation with which Mr. Sturgeon concluded our Visit. I was on the whole much pleas'd, and from what I then saw, have conceiv'd a higher Opinion of the natural Capacities of the black Race, than I had ever before entertained. Their Apprehension seems as quick,

> their Memory as strong, and their Docility in every Respect equal to that of white Children. You will wonder perhaps that I should ever doubt it, and I will not undertake to justify all my Prejudices, nor to account for them. (Franklin, 1763)

Franklin was thus a particular case of changing ideas on racism and slavery. From these different positions, nuances, and shifts in conviction within the United States' ruling class, a branch of the civil rights movement developed within the North American progressive bourgeoisie over time. In this way began the history of the struggle against discrimination, itself contradictory, torn, and never completely resolved—struggles which, as is well known, between 1861 and 1865 were among the principal causes of the American Civil War.

The distinctions and contradictions among Enlightenment thinkers on the racial question were therefore present in United States public debate. An example is provided by the vicissitudes linked to the Masonic values of liberty and equality, which, although intended in theory to extend to all humanity, were also applied only to Whites. In this milieu, a significant fact, seldom considered by historiography, was the recognition of the first fraternal societies composed of African Americans, such as Prince Hall Freemasonry, named after an African American Mason, ex-slave, and activist for Black education and the abolition of slavery. Prince Hall (c. 1735/8–1807) succeeded in having his African Lodge recognized by the Grand Lodge of England (an exceptional fact), and then emancipated it from England by founding an African American Grand Lodge active in Boston, Philadelphia, and Providence. These organizations, active since the last decades of the eighteenth century, formed the first African American leadership class, among whose members were many activists who, in the ensuing years, devoted themselves to changing North American society regarding discrimination and civil rights (Kantrowitz, 2010).

North American society, however, was composed of a severely fragmented bourgeoisie. Many intellectuals lived within these ambiguities regarding racism. For example, Albert Pike (1809–1891), an important Freemason at the head of the Scottish Rite in America, born in Boston in the North, nevertheless went to the Southern states to fight for the Confederates, and, after losing the Civil War, wrote explicitly racist texts, in some cases overlapping with segregationist ideas disseminated by the Ku Klux Klan.[2]

To highlight the importance of the differing perspectives on slavery and race within American Masonic brotherhoods, one can cite the debate surrounding the regulations that these fraternities accepted for the admission of members. Traditionally, they followed the rules indicated by James Anderson (c. 1679/1680–1739), a Scottish Protestant clergyman, reprinted precisely by Franklin in 1734, which specified as prerequisites for admission to the fraternities the qualities of being "good and true Men, free-born, ... no Bondmen" (Anderson & Franklin, 1734, p. 49). African American Masons had to fight to have recognized, as a qualification for admission into the brotherhoods, precisely the condition of being free and "not slaves," a category that allowed ex-slaves to participate with equal rights in white associational life. However, in Southern fraternities, where racism was a deeper problem, this possibility was readily eliminated (Hayden, 1866).

Along these lines, it must be recalled that activists for African American rights were, above all, formerly enslaved people, precisely like Prince Hall, and that they had a universalist worldview and, almost immediately, connected the abolition of slavery with the demand for women's citizenship rights. In this sense, Frederick Douglass (1818–1895) was a tremendously important figure, a former slave with a dramatic life story. He, in the years preceding and following the American Civil War (1861–1865), served as the principal point of reference for emancipationist groups, collaborating with Abraham Lincoln (1809–1865). In 1872, Douglass was presented as a candidate for the vice presidency alongside Victoria Woodhull (1838–1927), the first woman candidate for the presidency in United States history (who, however, at that time, as a woman, could not even vote for herself). Douglass united his activism for the civil rights of African Americans with struggles for women's emancipation. His name appears among the signatories of what is remembered as one of the first[3] feminist declarations in history:

[2] Following protests after George Floyd's 2020 killing, some statues of these North American white bourgeois figures with racist views were toppled. One of those removed was the statue of Pike in Washington.

[3] Olympe de Gouges (1748–1793), a playwright, wrote during the French Revolution (1789) a Declaration of the Rights of Woman and the Female Citizen (1791). This

> When, in the course of human events, it becomes necessary for one portion of the family of man to assume among the people of the earth a position different from that which they have hitherto occupied, but one to which the laws of nature and of nature's God entitle them, a decent respect to the opinions of mankind requires that they should declare the causes that impel them to such a course.
> We hold these truths to be self-evident; that all men and women are created equal; that they are endowed by their Creator with certain inalienable rights; that among these are life, liberty, and the pursuit of happiness; that to secure these rights governments are instituted, deriving their just powers from the consent of the governed. Whenever any form of Government becomes destructive of these ends, it is the right of those who suffer from it to refuse allegiance to it, and to insist upon the institution of a new government, laying its foundation on such principles, and organizing its powers in such form as to them shall seem most likely to effect their safety and happiness. [...].
> The history of mankind is a history of repeated injuries and usurpations on the part of man toward woman, having in direct object the establishment of an absolute tyranny over her. To prove this, let facts be submitted to a candid world.
> He has never permitted her to exercise her inalienable right to the elective franchise.
> He has compelled her to submit to laws, in the formation of which she had no voice.
> He has withheld from her rights which are given to the most ignorant and degraded men - both natives and foreigners.
> Having deprived her of this first right of a citizen, the elective franchise, thereby leaving her without representation in the halls of legislation, he has oppressed her on all sides.
> He has made her, if married, in the eye of the law, civilly dead. [...]. (Declaration of Sentiments, Seneca Falls Convention, 1848 in Stanton, 1889, p. 70)

Along these lines, it is essential to emphasize that the first feminist movement, which developed between the nineteenth and twentieth centuries,

feminist pioneer was anti-racist, abolitionist, and anti-colonialist, presenting works with incredibly avant-garde content. The Jacobins executed her due to her libertarian positions.

was far more significant than what is sometimes suggested in the somewhat naïve descriptions of the suffragettes often encountered.[4] The first feminists, organized in Europe by the Englishwoman Emmeline Pankhurst (1858–1928), carried out demonstrations of civil disobedience and militant antagonism (arson, hunger strikes, chaining themselves in protest) to obtain the vote, actions that triggered violent reactions from the institutions. In Italy, Maria Montessori, a scientist deeply engaged in the early feminist movement, was considered the prototype of the new woman[5]—a model that challenged the traditional power attributed to men. In this sense, Montessori's pedagogy was also "a militant science" in defense of women. Montessori first became concerned with women's rights, then with children's rights, and ultimately was engaged in international peace movements (Moretti, 2021).

The struggle of African Americans and the first wave of feminism thus shared the common objective of securing for formerly enslaved people and for women the inalienable rights already recognized for white men. The Fourteenth Amendment to the United States Constitution (1868) granted African Americans who had been enslaved the right to be considered persons and citizens with the same legal prerogatives as other citizens (the right to due process, to life, to liberty, and to property). In 1870, the right to vote was also recognized through the passage of the Fifteenth Amendment. Women, however, remained excluded, and white women did not obtain the right to vote until 1920 with the Nineteenth Amendment to the Constitution. We must wait until 1965, with the passage of the Voting Rights Act, for the extension of suffrage to all women and men in the United States.

[4] One example is Winifred Banks, a fictional Disney character and mother of the two protagonists in Mary Poppins, portrayed ridiculously as incapable of caring for her children due to her suffragist activism.

[5] Montessori was one of the first Italian women to graduate in medicine, focusing on hygiene, especially regarding perinatal behaviors of mothers, environmental health, and eventually pedagogy for first the abnormal and later all children. In an initial phase, Montessori engaged in craniometric measurements of children, trying to correlate skull size with intellectual evaluation; her anthropological research was inspired by Léonce Manouvrier (1850–1927), known for proposing measurements of female skull volume that, when related to the whole body, would demonstrate female brain superiority over males. Manouvrier's anthropology was alternative to Broca's or Lombroso's, chiefly because it posited that bodily measurements could not be signs of degeneration or immorality; instead, anti-social behavior was more the product of an individual's environment, which in turn influenced bodily dimensions.

The trajectory of this history is interesting because it reveals an internal cyclicality: whenever one of the racialized groups organized and achieved results on the political level, a "counter-reaction" arose from those who felt they were losing power relative to those who were gaining it. Thus, after the abolition of slavery between 1863 and 1865 by Lincoln, racists reacted violently, and the first victim of this reaction was the American president himself, killed by a Southerner frustrated by the outcomes of the Civil War. Abolitionism and emancipation, as is well known, were opposed by white supremacist groups such as the Ku Klux Klan who, at least in the Southern states—even after losing the war—succeeded in reintroducing segregationist laws restricting the personal liberty of formerly enslaved people (Jim Crow Laws), which, as we shall see, were dismantled only with difficulty and only in the course of the twentieth century. Another example of this historical cyclicality: Fascism and Nazism themselves can be considered ultraconservative movements reacting against liberal modernity, which was emerging powerfully in the early decades of the twentieth century, staging a tragic anti-progressive revolt that led to the destruction of the Second World War.

In contemporary history, the Enlightenment values, the new forms of constitutional government, and the struggles of workers, abolitionists, and emancipationists had instead promoted significant cultural changes in the recognition of the rights of workers, African Americans, and women; yet in all these cases, there were always conservative counterparts opposing them. In this sense, the Jews became the scapegoats of resistance to change because they were identified as the industrious people who were modernizing the world. Modernization was viewed, by the dominant groups, not as a resource but as a danger (Foschi, 2024).

Fundamental to the deconstruction of racism and prejudice was therefore the presence of marginalized and racialized groups among the ranks of activists. In the same way, also in science, those who advanced original, non-conservative positions were Jews, African Americans, and women. These researchers had a distinctive inclination toward debunking racist theories.

Interesting from this perspective is the story of Franz Boas (1858–1942), a Jewish physicist who, between the end of the nineteenth and the beginning of the twentieth century, became one of the foremost anthropologists. Born in Germany, he emigrated to the United States in the late nineteenth century. Boas rejected the paradigm that culture derived from biology, and he was convinced that they should be considered on separate planes, without making culture dependent on genetics. First of all, he demonstrated that even physical elements such as measurements of the skeleton and cranium depended on nutrition and thus on environmental factors. Boas, subsequently, gave autonomy to cultural studies, inaugurating a new approach to anthropology. By employing Galtonian statistical methods and demonstrating that certain physical traits in different environments, whether richer or poorer, were not stable across generations, he questioned the linearity of progressive development, which evolutionists believed in. From one generation to the following, significant modifications may emerge due to environmental variations or unforeseen dynamics. Although some of Boas's research has been criticized, it is remarkable that this scholar dared to move against the mainstream in an era in which hard inheritance was gaining ground (see Chapter 5). Boas also assisted dissident scientists during Nazism; his was a deliberately militant and anti-racist science (Kuznick, 2019). He also supported the first African American intellectuals who began to make the racial question central to their scholarly work. Boas had, for example, an enormous impact on William E. B. Du Bois (1868–1963), an African American sociologist and activist for the rights of African Americans (Liss, 1998). Thus did Du Bois describe the encounter between Boas and his students:

> Franz Boas came to Atlanta University where I was teaching history in 1906 and said to a graduating class: You need not be ashamed of your African past; and then he recounted the history of the black kingdoms south of the Sahara for a thousand years. I was too astonished to speak. All of this I had never heard and I came then and afterwards to realize how the silence and neglect of science can let truth utterly disappear or even be unconsciously distorted. (Du Bois, 1939, p. VII)

The students of Boas, moreover, took up the issue of the eugenic use that was made in North America of the Binet test (Cicciola, 2019). Otto Klineberg (1899–1992), a white psychologist of Jewish origin, demonstrated that some environmental and cultural factors, more or less recent, influenced differences in intelligence quotient both among ethnic groups and within the same ethnic groups. Along these lines, African American researchers such as Horace Mann Bond (1904–1972) demonstrated that African Americans in the North had average intelligence test scores superior to those of Southern Anglo-Saxons, and that this was due to contextual factors, rather than biological ones. Slavery had ended many decades earlier, but the damage inflicted upon families and generations by racism, prejudice, and the segregationism that replaced slavery was still very much present. In this sense, the work of Gunnar Myrdal (1898–1987) in the United States was fundamental. Myrdal was a controversial social democratic economist from Sweden, who in his homeland defended the eugenic program of sterilization of the feeble-minded (Spektorowski & Mizrachi, 2004). In the United States, he criticized the treatment reserved for African Americans, racism, which he considered an actual American pathology. Already in the 1940s, based on much data collected in research financed by the Carnegie Corporation, Myrdal demonstrated that the United States was not living up to the founding ideals on which it was supposedly based. The condition of African Americans was miserable and unresolved; segregation was unworthy. It worsened an already precarious situation, provoking a reaction among Blacks who, marginalized, were in turn increasingly stigmatized by Whites (Myrdal, 1944). Segregation, with a vicious circle, further burdened the social conditions of African Americans (for example, separating Blacks and Whites in hospitals and schools, which worsened conditions also from the point of view of institutional and economic organization in segregationist states); moreover, immediately after the Second World War, African Americans had become suspicious of and frightened by the possibility of their cultural assimilation. Myrdal's solution to the American racist pathology was precisely a kind of assimilation of Black culture; it was obviously much criticized by activists who instead wanted African American culture to be preserved and valued.

Myrdal's studies were widely circulated and discussed among civil rights activists in the 1950s and 1960s. At that point, those who had already put into question racial studies moved on to investigate the prejudices and stereotypes underlying all racist psychological theories. It was revealed that white children grew up assimilating stereotypes about Blacks, while Blacks tended to avoid intermixing with Whites. In particular, it is necessary to recall the studies of the couple Kenneth (1914–2005)[6] and Mamie Clark (1917–1983), African American psychologists, who demonstrated through a famous experiment that African American

[6] Kenneth was the first African American president of the American Psychological Association in 1966. In reality, after Clark, the APA has seen an increase in African American and female presidents only in very recent years.

children between the ages of three and seven preferred to play with white dolls rather than black ones. This revealed a profound lack of self-esteem caused by the racist context in which these children were growing up, believing themselves to belong to an ethnicity with "inferior" characteristics (Clark & Clark, 1950). These studies led the United States Supreme Court to take a historic decision in 1954, abolishing school segregation.

In the field of anti-racist psychological studies, it is also necessary to remember the many Jewish scholars who arrived in the United States due to persecution.[7] In particular, the leading exponents of the Frankfurt School, who undertook a research program precisely to understand the culture that in Europe had produced racism and Nazifascism. Fundamental in this regard was the contribution of Theodor W. Adorno (1903–1969) who, having escaped the Nazis and emigrated to the United States, coordinated studies that led to the definition of authoritarianism and to the construction of the famous E- and F-scales, which measured various dimensions of the ethnocentric, antisemitic, authoritarian, and fascist personality (See Adorno et al., 1950). In this group, Else Frenkel-Brunswik (1908–1958) began to study cognitive intolerance, which produced polarizations (black-white, good-evil, right-wrong), an authoritarian mentality connected to the logic typical of racism (Montuori, 2005; see also Lauriola et al., 2015).

Building upon these studies, Stanley Milgram (1933–1984) was destined, even within the span of a short life, to play an important role in the history of psychology. A Jew raised in a family persecuted by Nazism, Milgram studied the phenomenon of obedience to authority. He devised a series of psychological experiments to understand why the Shoah had occurred. Between 1960 and 1963, at Yale University's Department of Psychology, he conducted a research program initially publicized as a study on memory and learning. Those who volunteered were told that they would be randomly assigned to either the role of "teacher" or "learner." Each participant was paid $4 per hour. In reality, by means of a setup, the "learner" was always an accomplice of the experimenter, while the "teacher" was the actual participant, unaware of his real experimental purpose. The "teacher" had to ask the learner to associate a word correctly with another word on a list that the learner was supposed to

[7] Recently, Patrizia Guarnieri of the University of Florence reconstructed the networks that helped European Jews escape Nazism by emigrating to the United States. See the website intellettualinfuga.com.

have previously memorized. The system presupposed that the learner was in another room, strapped to electrodes. The unsuspecting teacher was led to believe that the aim of the experimenter was to study the effectiveness of punishment on learning. After an incorrect answer, the "teacher" was required to administer an electric shock of increasing intensity using a "shocking generator," which was fictitious but seemed real to the participant. The learner communicated with the teacher through an intercom, and the maximum shock to be administered was said to be of a value that could be dangerous to life. When the learner answered correctly, the teacher was to congratulate him and proceed with the test; when wrong, administer a shock.

The "real" experiment consisted of observing to what extent the "teacher" would continue to administer increasingly stronger shocks even as he heard the learner in the other room complaining more and more urgently, until finally, after the umpteenth (fictitious) shock, he heard no further sound. In the teacher's room, the researcher, representing the authority to be obeyed, instructed the teacher to continue the experiment, even as the shocks seemed to cause real pain to the learner next door.

Milgram had previously asked a large group of psychiatrists to predict the behavior of the participants, and almost none judged it possible that teachers would go as far as administering the maximum voltage. In contrast, two participants out of three reached the maximum shock. The most interesting aspect of Milgram's various trials, however, concerned the situational factors that favored either compliance and the administering of maximum shocks, or rebellion against the experimenter and abandonment of the experiment. By varying the context, if the teacher was placed within a team of other "sadistic" teachers (all accomplices) who administered shocks in his place, almost no teachers rebelled, and more than 90% completed the shocks to the end (experiment 18). Conversely, when teachers—again, accomplices—were shown openly rebelling against the experimenter, the subject-participant likewise tended to rebel, and only 10% ultimately administered the maximum shocks (Experiment 17) (Milgram, 1974)

Milgram's social experiment attracted both admirers and critics. To understand its importance, however, it must be contextualized within an anti-racist and antiauthoritarian interpretive framework. Milgram used, in fact, the experimental social psychology of his time to attempt to uncover the psychological factors that led to passive acceptance of authority and contributed to behaviors similar to those of the Nazis, ordinary people who obeyed uncritically. What emerged was a dark picture, but one mitigated by the possibility of altering contexts in such a way that people might become aware and avoid the aberrations of their own behavior, identifying with and interacting with virtuous models of behavior.

In the postwar period, Ashley Montagu (1905–1999), an English cultural anthropologist and anthropometrist of Jewish origin, was probably the most significant figure in the history of anti-racism during the Cold War. A student of Boas, Montagu deconstructed the myth of race with empirical studies and logical observations that remain foundational to critical thought on racism. He authored a famous volume about race in which even the title expressed the core concept: *Man's Most Dangerous Myth: The Fallacy of Race*:

> The idea of "'race'' represents one of the most dangerous myths of our time, and one of the most tragic. Myths are most effective and dangerous when they remain unrecognized for what they are. Many of us are happy in the complacent belief that myths are what primitive people believe in, but that we ourselves are completely free of them. We may realize that a myth is a faulty explanation leading to social delusion and error, but we do not usually realize that we ourselves share in the mythmaking faculty with all men of all times and places, that each of us has his own store of myths which have been derived from the traditional stock of the society in which we live. In earlier days we believed in magic, possession, and exorcism, in good and evil supernatural powers, and until recently we believed in witchcraft. Today many of us believe in "race." "Race" is the witchcraft of our time. The means by which we exorcise demons. It is the contemporary myth. Man's most dangerous myth. (Montagu, 1952, p. 1)

Montagu's analysis was precise and aimed to demonstrate that race is a false concept from which no moral or psychological qualities, inferior or superior, can be derived. In this sense, he served as coordinator of the scholars who authored the first famous "statement" of UNESCO (1950). For the first time (though with difficulty), the statement sought to deny both the concept of race and the existence of any scientifically proven link between race and specific behaviors or personalities:

> To most people, a race is any group of people whom they choose to describe as a race. Thus, many national, religious, geographic, linguistic or

cultural groups have, in such loose usage, been called "race" when obviously Americans are not a race, nor are Englishmen, nor Frenchmen, nor any other national group. […] National, religious, geographic, linguistic and cultural groups do not necessarily coincide with racial groups: and the cultural traits of such groups have no demonstrated genetic connexion with racial traits. Because serious errors of this kind are habitually committed when the term "race" is used in popular parlance, it would be better when speaking of human races to drop the term "race" altogether and speak of ethnic groups. […] Whatever classification the anthropologist makes of man, he never includes mental characteristics as part of those classifications. It is now generally recognised that intelligence tests do not in themselves enable us to differentiate safely between what is due to innate capacity and what is the result of environmental influences, training and education. […] The scientific material available to us at present does not justify the conclusion that inherited genetic differences are a major factor in producing the differences between the cultures and cultural achievements of different peoples or groups. It does indicate, however, that the history of the cultural experience which each group has undergone is the major factor in explaining such differences. The one trait which above all others has been at a premium in the evolution of men's mental characters has been educability, plasticity. This is a trait which all human beings possess. […] So far as temperament is concerned, there is no definite evidence that there exist inborn differences between human groups. […] As for personality and character, these may be considered raceless. In every human group a rich variety of personality and character types will be found, and there is no reason for believing that any human group is richer than any other in these respects. With respect to race-mixture, the evidence points unequivocally to the fact that […] There is […] no biological justification for prohibiting intermarriage between persons of different ethnic groups. […] "Race" is not so much a biological phenomenon as a social myth. The myth of "race" has created an enormous amount of human and social damage. In recent years it has taken a heavy toll in human lives and caused untold suffering. It still prevents the normal development of millions of human beings and deprives civilization of the effective co-operation of productive minds. […] The unity of mankind from both the biological and social viewpoints is the main thing. To recognize this and to act accordingly is the first requirement of modern man. […] And, indeed, the whole of human history shows that a cooperative spirit is not only natural to men, but more deeply rooted than any self-seeking tendencies. […] We now have to consider the bearing of these statements on the problem of human equality. It must be asserted with the utmost emphasis that equality as an ethical principle in no way depends upon the assertion that human

> beings are in fact equal in endowment. Obviously individuals in all ethnic groups vary greatly among themselves in endowment. Nevertheless, the characteristics in which human groups differ from one another are often exaggerated and used as a basis for questioning the validity of equality in the ethical sense. [...] Historical and sociological studies support the view that genetic differences are not of importance in determining the social and cultural differences between different groups of *Homo sapiens* [...]. Vast social changes have occurred which were not in any way connected with changes in racial type. [...] All normal human beings are capable of learning to share in a common life, to understand the nature of mutual service and reciprocity, and to respect social obligations and contracts. Such biological differences as exist between members of different ethnic groups have no relevance to problems of social and political organization, moral life and communication between human beings. Lastly, biological studies lend support to the ethic of universal brotherhood; for man is born with drives toward co-operation, and unless these drives are satisfied, men and nations alike fall ill. Man is born a social being who can reach his fullest development only through interaction with his fellows. The denial at any point of this social bond between man and man brings with it disintegration. In this sense, every man is his brother's keeper. For every man is a piece of the continent, a part of the main, because he is involved in mankind. (UNESCO, 1950, pp. 6–10)

This selection from the document is representative of all those passages that were particularly innovative compared to what had previously been written about race. The text, as these extracts show, was also extremely heterogeneous and, among other things, did not entirely exclude the possibility of biological ethnic differences, but emphasized that such biological differences were not important. In the following years, the 1950 statement was modified and revised, first in a biologistic and then in a culturalist sense. Montagu, above all, continued to be a reference point also for the later statements on the subject.

After the atrocities committed by "scientists" espousing racist eugenics, geneticists too developed a strong sense of Pride in scientific research. Thus, especially after the Second World War, there emerged among them universal positions of condemnation of fascist and Nazi racism, even while variable positions remained with respect to genetics itself, producing a crisis within the discipline. Some have subjected hard inheritance to empirical scrutiny, thereby producing research that is more attentive to facts and population dynamics. Others opposed hard inheritance with an equally hard and ideological environmentalism, a doctrine particularly developed in the Soviet Union based on the work of the agronomist-biologist Trofim D. Lysenko (1898–1976). Still others rejected racial typologies but not the importance of genetics in determining traits and individual differences; from this perspective, the more progressive geneticists believed that only egalitarian policies should moderate the individual differences stemming from genetics (Jackson & Weidman, 2006).

In this sense, UNESCO's 1950 statement was revised for the first time the following year by a group of geneticists, whose work Montagu also participated in. The second text, *Statement on the Nature of Race and Race Differences* (1951), appears to have been revised and is significantly more moderate. The emphasis on human fraternity was eliminated, as was the idea that race is a myth. It insists that there is no evidence of genetic differences in intelligence or personality, but does not exclude the possibility a priori (UNESCO, 1969, pp. 36–43). This text too was criticized, especially by Galton's English academic descendants, such as Ronald Fisher, who were combative in interpreting intellectual and personality differences evidenced by testing as the result of genetic differences rather than methodological errors or cultural variations. Galton's heirs continued to be fascinated by eugenics and to hold racist attitudes. UNESCO's declarations generated much discussion, and scientists presented differing opinions at least until 1978, when the last version of these declarations (the fifth), *Declaration on Race and Racial Prejudice*, was produced. This text eliminated every conception of the existence of human races and of biological priority in explaining differences among humans (thus returning to, and in fact radicalizing, the 1950 declaration). The first two articles read:

Article I

1. All human beings belong to a single species and are descended from a common stock. They are born equal in dignity and rights and all form an integral part of humanity.
2. All individuals and groups have the right to be different, to consider themselves as different and to be regarded as such. However, the diversity of life styles and the right to be different may not, in any circumstances, serve as a pretext for racial prejudice; they may not justify either in law or in fact any discriminatory practice whatsoever, nor provide a ground for the policy of apartheid, which is the extreme form of racism.

3. Identity of origin in no way affects the fact that human beings can and may live differently, nor does it preclude the existence of differences based on cultural, environmental and historical diversity nor the right to maintain cultural identity.
4. All peoples of the world possess equal faculties for attaining the highest level in intellectual, technical, social, economic, cultural and political development.
5. The differences between the achievements of the different peoples are entirely attributable to geographical, historical, political, economic, social and cultural factors. Such differences can in no case serve as a pretext for any rank-ordered classification of nations or peoples.

Article 2

1. Any theory which involves the claim that racial or ethnic groups are inherently superior or inferior, thus implying that some would be entitled to dominate or eliminate others, presumed to be inferior, or which bases value judgements on racial differentiation, has no scientific foundation and is contrary to the moral and ethical principles of humanity.
2. Racism includes racist ideologies, prejudiced attitudes, discriminatory behaviour, structural arrangements and institutionalized practices resulting in racial inequality as well as the fallacious notion that discriminatory relations between groups are morally and scientifically justifiable; it is reflected in discriminatory provisions in legislation or regulations and discriminatory practices as well as in anti-social beliefs and acts; it hinders the development of its victims, perverts those who practice it, divides nations internally, impedes international co-operation and gives rise to political tensions between peoples; it is contrary to the fundamental principles of international law and, consequently, seriously disturbs inter- national peace and security.
3. Racial prejudice, historically linked with inequalities in power, reinforced by economic and social differences between individuals and groups, and still seeking today to justify such inequalities, is totally without justification (UNESCO, 1979, pp. 11–13).

That said, it is essential to note that between the 1950s and 1960s, the civil rights movement underwent significant differentiation. Activists opposed to the notion that African American culture was irreparably inscribed in a pathological vicious circle, based on Myrdal's ideas, developed more radical positions in opposition to white power. Many episodes of open conflict between Blacks and Whites occurred. The Black Power movement was born, tied precisely to the concept of Black "Pride." The intellectual who most influenced these positions was Frantz Fanon (1925–1961), a French psychiatrist born in the French colony of Martinique, a critic of colonialism who studied the psychopathological damage caused by the state of cultural subordination experienced by dominated peoples.

In the meantime, the year 1963 was key to the abolition of segregationist laws. One hundred years after Lincoln's first abolitionist proclamation, on August 28, 1963, Martin Luther King Jr. (1929–1968) marched for jobs and freedom in Washington, on the steps of the Lincoln Memorial, and delivered the famous "I Have a Dream" speech before hundreds of thousands, calling for an end to racism and segregationism. The famous March on Washington was an event that became iconic for all subsequent protest marches. The creator and main organizer of the march was Bayard Rustin (1912–1987), an African American activist and gay man with a history of involvement with the Communist Party. Rustin thus embodied different identities that were subject to discrimination and was therefore a sort of prototype of intersectionality as it is understood today. These different identities, which did not fit perfectly into the mid-1960s civil rights movement, meant that he was largely forgotten until recently.[8]

On June 19, 1963, President John F. Kennedy (1917–1963) presented the Civil Rights Act proposal, which effectively abolished segregation in schools and employment for African Americans. The law was passed the following year, but in the meantime, the reaction to this progressive politics had led to Kennedy's assassination, in a situation that appeared strikingly similar to that of Lincoln.[9] The entire anti-racist and egalitarian movement, supported by the more liberal political class, instilled

[8] In 2013, President Obama remembered Rustin by posthumously awarding him the Presidential Medal of Freedom, describing him as "the angelic troublemakers every community needs."

[9] J.F.K. was assassinated on November 22, 1963, in Dallas under circumstances that remain unclear. Recently declassified documents point to the involvement of the most conservative factions of the American secret services, who opposed both the domestic

fear among the reactionary elements of American society; this resulted in the assassination of the most prominent progressive leaders: Martin Luther King and Robert Kennedy (1925–1968), brother of John and Democratic candidate in the presidential primaries following his brother, who were both killed in 1968.[10]

In the following years, Fanon's critiques entered the cultural repertoire of 1960s activists, culminating in the French May of'68 and the demonstrations that erupted throughout much of the West that year. The movement of 1968 was that of demonstrations for civil rights, for greater sexual freedom, of the second wave of feminism, and for the definitive liberation of homosexuals. The liberation movements proclaimed self-determination on issues such as contraception, abortion, divorce, premarital relations, equal rights in the workplace, etc. Pietro Clemente, an Italian anthropologist, recalls the influence of Fanon[11] on the "movement":

> At the origin of this research lies the suggestion that Frantz Fanon had an impact on this writer, as well as on the entire Youth and student movement that exploded in'68. In that context, more than with the events of the Algerian War, Fanon connected with the struggles of African Americans, with Vietnam, with Che Guevara, creating the composite but exalting framework of names and events which, together with Valle Giulia and the Paris May, appeared as both proofs and symbols of the resumption of a revolutionary and violent struggle, of the political project of peaceful competition and of the control of the traditional formations of the workers' movement. Fanon stood within this framework as the prophet of the violence of the oppressed, the forerunner of a culture totally "other" from that of white capitalism, a supporter, therefore, of the radical rejection of the "West". (Clemente, 1971, p. 5)

Fanon especially influenced intellectuals such as Angela Davis, who embodied the more radical wing of the African American civil rights

and international policies of the Kennedy family. See Oliver Stone's movie *JFK Revised* (2021).

[10] It is also well known that Malcolm X (1925–1965), another Black Power leader, was assassinated in 1965.

[11] Fanon's classic work, *Les Damnés de la Terre* (The Wretched of the Earth), published in 1961, was an immediate international success; in Italy, for example, it was translated as early as 1962.

movement, introducing the Marxist critique[12] into the U.S. debate, and Davis integrated feminist struggles with anti-racist efforts. Between the 1960s and 1970s, the most radical civil rights movement attempted to unite different marginalized cultures and groups, bringing together African Americans, Hispanics, women, homosexuals, and the working class to fight for the same goals. An example of this type of social struggle was the Rainbow Coalition, active in Chicago until the early 1970s, which the FBI then dismantled due to its revolutionary and radical nature. However, a few years later, the idea of a Rainbow Coalition that would unite different causes resurfaced in a more pacifist context and was organized by Jesse Jackson, heir to Martin Luther King, a Baptist minister and member of the Prince Hall Masonic Lodge. Jackson, a former Democratic presidential candidate in the 1980s, remains a leader of the civil rights movement.

About the movement for homosexual rights, everything began around the same period with the Stonewall riots in New York. The incessant police raids in bars where gays gathered were met with violent resistance between June 28 and 29, 1969, at the Stonewall Inn in Greenwich Village, New York. The spark erupted when trans people, led by Sylvia Rivera (1951–2002), threw objects at the police, starting and leading the clashes. Thus was born Gay Power, explicitly modeled on the African American and feminist struggles. One year after the Stonewall events, the American Gay Liberation Front inaugurated Gay Pride in various American cities, which continues today as the LGBTQ+ Pride Parade, held precisely around the dates commemorating Stonewall; Pride, or the parade of Pride, is thus an attempt to heal the wounds of discrimination and racisms. The rainbow flag, symbol of peace, was adopted as the emblem of the movement. In 1972, Pride was held for the first time in Europe, in London. In various nations, homosexual liberation organizations were founded, actively participating in the broader movements of the 1970s (Faderman, 2015). The LGBTQ+ movement has been a significant force in our history, mainly due to its efforts in contemporary identity politics, which aim to unite civil rights claims based on intersectionality and challenge binary thinking (De Leo, 2021).

[12] The Black Power Movement included various factions, ranging from the more moderate and pacifist to the more revolutionary. Marxism primarily entered through the Black Panther Party, which eventually adopted Marxist-Leninist positions, ultimately leading to its demise at the hands of the CIA and FBI.

Nevertheless, homosexuals continued to be considered sociopaths, inferiors, deviants, and abnormal. Psychiatry continued to support this view, and although Freud had foreshadowed the possibility of a dynamic sexuality, classical psychoanalysis, that of the official institutes, maintained a conservative and "psychopathological" stance toward homosexuals. Prejudice abounded even in progressive circles. In 1972, a psychiatrist, Dr. John Fryer (1937–2003), presented himself in mask and wig as Dr. Henry Anonymous at the annual convention of the American Psychiatric Association in Dallas, delivering a historic speech while using a voice distorter:

> I am a homosexual. I am a psychiatrist. I, like most of you in this room, am a member of the APA and am proud to be a member. However, tonight I am, insofar as in it is possible, a "we." I attempt tonight to speak for many of my fellow gay members of the APA as well as for myself. [...] I am disguised tonight in order that I might speak freely without conjuring up too much regard on your part about the particular who I happen to be. [...].
> As psychiatrists who are homosexual, we must know our place in what we must do to be successful. If our goal is academic appointment, a level of earning capacity equal to our fellows or admission to a psychoanalytic institute, we must make certain that no one in a position of power is aware of our sexual orientation or gender identity. Much like the black man with the light skin who chooses to live as a white man, we cannot be seen with our real friends - our real homosexual family - lest our secret be known and our dooms sealed. There are practicing psychoanalysts among us who have completed the training analysis without mentioning their homosexuality to their analysts. Those who are willing to speak up openly will do so only if they have nothing to lose, then they won't be listened to. [...].
> When fellow homosexuals come to you for treatment, don't let your own problems get in your way, but develop creative ways to let the patient know that they're all right. And teach them everything they need to know. [...]. (Fryer, 1985, pp. 1–3)

Gay Pride and activism thus led to the elimination of homosexuality as an actual pathological deviation in the seventh edition of the DSM-II (1974).[13] In the 1970s, therefore, all the conditions were in place for

[13] The reference to homosexuality, however, remained for a long time under various labels such as egodystonia, a mental distress caused by feeling one's gender identity

Western society to undergo a radical transformation compared to its situation at the beginning of the twentieth century. Women could finally obtain the rights that had been denied them and for which they had fought for over a century, and the same was true for homosexuals. As we have seen, the protest and civil rights movements of the late 1960s called for a radical change in values toward pacifism, egalitarianism, anti-racism, and feminism. After the Stonewall riots (1969), the LGBTQ+ rights movement also developed rapidly. According to Inglehart and Norris (2016), the twentieth century in advanced capitalist countries witnessed a revolutionary shift in customs, accompanied by a reallocation of power that was experienced as a crisis by conservative social classes (see also Norris & Inglehart, 2019). A sort of revolution within the context of the Cold War (on the silent revolution, see Inglehart, 1977).

Consequently, these changes were perceived as negative by a large part of the population. Thus, against these revolutions, conservative reactions always emerged, aimed at blocking change out of fear of "disorder" and "chaos," sentiments with which traditionalists often experienced social transformation (Foschi, 2024).[14] To understand the arguments of those fearful of change, we can follow the reasoning of Evola—whom we have already encountered several times—as a dystopian megaphone for these anxieties and fears of change. He wrote numerous essays in the 1960s and 1970s in which he dismissed all the contemporary transformations, addressing precisely the issues we have thus far highlighted. Evola was utterly opposed to the emancipation of African Americans and women, to the 1968 movement, and to the movement for sexual liberation. He was equally opposed to the emergence of the Fifth Estate (a new social class composed of young people engaged in creative work, which, after the clergy, the aristocracy, the bourgeoisie, and the proletariat, was considered characteristic of advanced capitalist societies).[15] Evola called for

different from biological sex. However, homosexuality itself was no longer considered a deviation or psychopathological category.

[14] This is the phenomenon known in social psychology as System Justification Theory: the tendency of some people to preserve the social structure for fear of change, which they denigrate, along with those who promote it. For further information, see Jost (2020).

[15] Recently, The Fifth Estate has been referred to as the class of atypical workers, especially young, creative, and knowledge workers (cf. Allegri & Ciccarelli, 2013). Evola, on the other hand, considered the Fifth Estate to be mainly the young people of the counterculture of the 1960s and 1970s. However, it is precisely these young people who represent the matrix from which the current Fifth Estate has developed.

revolt against the modern world, against Africanization, against feminism, against new forms of sexuality, against "ethnic replacement." Therefore, the relationship with tradition in the thinking of traditionalists such as Evola can easily become nostalgia; in the worst case, it can lead to a rejection of all forms of progress and modernity, even to the point of being blocked by force (Foschi, 2024). The following are some significant passages on these topics.

Against African Americans:

> Not long ago, newspapers announced that, according to some calculations, by 1970, half of Manhattan's population would be black and that in the five boroughs that make up the entire city of New York, 28% of the inhabitants would be of the black race. Similar developments have been observed in other cities and regions of the United States. We are witnessing a negrification, a mongrelization, and a decline of the white race in the face of inferior races that reproduce more rapidly. [...] It seems that it is precisely the Negro element, in its primitivism, that has set the tone for important aspects of the American psyche. It is already significant that when we talk about American folklore, we refer to Negroes as if they were the original inhabitants of the country. Thus, the famous *Porgy and Bess*, by the Jewish composer Gershwin, which deals exclusively with Negroes, is considered in the United States to be a classic work inspired by "American folklore." The composer stated that he had lived among American Negroes for some time in preparation for this work. But the phenomenon of popular and dance music is even more evident and widespread. Fitzgerald was not wrong when he said that, in one of its main aspects, American civilization can be defined as a civilization of jazz, that is, of Negroized and Negroizing music and dance. In this field, very singular "elective affinities" have led America, through a process of regression and primitivization, to imitate Negroes [...]. (Evola, 1968–2000, pp. 39–41)

Against Youth Movements:

> Much, indeed, too much, has been written on the subject of the new generation and "young people." In most cases, the topic does not deserve the attention it has received. The importance sometimes attributed to Youth in general today [...]. When these young people claim to be misunderstood, the only answer that can be given to them is that there is simply nothing to understand about them — that, if a normal order were in force, it would suffice to put them in their place by force, as one does with children when their stupidity becomes annoying, intrusive, and impertinent

[...]. Many of the phenomena we have examined in the preceding pages, such as the taste for vulgarity and certain new social customs, can be largely attributed to this Youth. Examples include the fanatics (male and female) of the screamers—those epileptic "folk singers"—and, currently, of the collective puppet show known as "ye-ye sessions" and of this or that album, with all that such interests entail in terms of behavior. Their total lack of a sense of the ridiculous makes it impossible to exert any influence on them, so all that remains is to leave them to their fate and their stupidity. (Evola, 1968–2000, pp. 191–192)

Against Women's Emancipation and Sexual Freedom:

Certain sectors of contemporary society are reaching a climate of widespread and chronic sensuality, through a constant and insidious appeal to men by sex and women, which leads to the decay of all superior masculine values and true spirituality, in line with what has always happened in the final and twilight phases of many cycles of civilization. Indiscriminate sexual freedom is all the more dangerous for a people such as the Italians, who, due to their race rather than simply as a consequence of traditional prejudices, certainly do not have the best disposition in the sexual sphere. For example, apart from the more sophisticated types, young Italian women find it almost impossible to go beyond the alternative of being either a silly goose or vulgar. It is easy to talk about economic emancipation and sexual freedom as an antidote to prostitution: in reality, the only case in which this freedom does not prove harmful is when women have acquired a special personality, something they have not even begun to do today, despite all the achievements and claims of women on a material and practical level. In Italy, the last war only led to an increase in petty, trivial, or mercenary corruption. A recent survey of Italian prostitutes found that the prevailing view among them is no different from that of the *petite bourgeoisie* girls who are looking for a husband and eager to "settle down": it is by no means the view of young women who aspire to greater freedom and nonconformity and who make use of it. (Evola, 1968/2000, p. 134)

Against Homosexuality:

It may be thought that by regression this dominant power upon which sexualization depends—the being truly man or woman—weakens. Then, just as politically in society, when every central authority grows weak, all the forces from below, previously restrained, can be released, so too in the individual, the latent characteristics of the opposite sex can surface, and thus a tendency toward bisexuality. One then finds oneself faced with the condition of a "third sex," and it is obvious that here we have particularly fertile ground for the phenomenon of homosexuality. The premise is thus an internal capitulation, a waning of the "inner form," or, more precisely, of the formative power, which manifests itself not only in sex but also in character, in personality, in the possession, in general, of a "definite character." It becomes understandable, therefore, why the aforementioned development of homosexuality in endemic forms is a sign of the times, a phenomenon which logically belongs to those through which the modern world presents itself as a regressive world. (Evola, 1997, pp. 329–330)

Against the Fifth Estate:

The decline through history would not stop, then, at the Fourth estate, that is, at the collectivized Marxist and communist world; rather, it would continue in the emergence of a Fifth estate. What would this Fifth estate be? Here, it is necessary to refer above all to the idea that every organization comprises two elementary principles: forces of order on the one side, forces of chaos on the other. Each organization arises from a formative action that binds and restrains, within specific structures (where they can manifest themselves creatively as a dynamic factor), the latter. Well, when a cycle draws to its end, this elementary substratum, the subpersonal foundation—almost one could say in Goethean terms "demonic"—which in traditional civilizations had been bent, restrained, and elevated by a higher law and by the natural prestige that spiritual, heroic, and aristocratic values and their representatives possessed, tends to return to its free state, to act destructively, to prevail. This is the limit, and this may correspond to the advent of the Fifth estate [...]. One might, for example, mention certain aspects of the so-called "generation in revolt." Revolt can be legitimate when directed against a civilization in which almost nothing retains any higher justification, a civilization that is empty and absurd, which, mechanized and standardized, itself tends toward the subpersonal in an amorphous world of quantity. But when one encounters "rebels without a flag," when revolt is, so to speak, an end in itself, the rest serving as a pretext, when it is accompanied by forms of unchaining, primitivism, abandonment to what is elemental in an inferior sense (sex, Negro jazz, intoxication, gratuitous and often criminal violence, complacent exaltation of the vulgar and the anarchic), then it is not hazardous to establish a nexus between these phenomena and others which, on another plane, attest to the action of chaotic forces rising from below through the increasingly visible fissures of the extant order, forces by which are possessed those elements cast adrift and more or less traumatized. (Evola, 1969, pp. 684–689)

Despite his racism and the manifest tendence to discrimination, Evola's thought is still today considered a source of inspiration and legitimation for currents of the "new" right, both moderate and conservative, a source of inspiration also for American (Bannon) and Russian (Dugin) political theorists who today are fighting against the modern world for the restoration of an idealized past (Foschi, 2024; Lachman, 2018).

If Evola represented a theoretical voice of reaction to the "silent revolution," there was also a reaction from those "scientists" particularly attached to pseudo-biological concepts, so that the undermining of the concept of race and the critique of eugenics also triggered the response of the last heirs of Galton. This group of particularly combative mainstream psychologists, engaged in the study of individual differences and psychometrics, especially in England, defended with vigor the idea that some people had a lower intelligence quotient, as measured through questionnaires and correlations, and their misfortunes and poverty were caused by low IQ. The actual influence of Galton's brainchild upon psychology, therefore, always deserves to be recalled, critically and in depth. Notable names in contemporary scientific research in psychology, such as Hans Eysenck (1916–1997), identified with this approach, which appeared well rooted in biology, genetics, and the objective measurement of behavior. The horizon of these behavioral scientists was to provide expertise for reforms in democratic states by virtue of a presumed efficiency and eugenic rationalization, exactly as in Galton's novella *Kantsaywhere*. Institutions performed their part by conducting evaluations and administering tests to the population. Even liberal organizations, such as the American Psychological Association (APA), were influenced by the eugenic conception held by many of its prestigious members. Yerkes and Terman, for instance, were even presidents of the APA. As we have seen, Fisher was critical of UNESCO's declarations on race.

Even later, when biologists became increasingly confident that the notion of race for humans was pseudoscientific for empirically demonstrable reasons, psychologists of the English psychometric tradition continued to stubbornly oppose the abandonment of the category. The concept of heritability of intelligence or personality has remained a debated topic in psychology, one that retains its adherents and, in any case, tends to recur cyclically, just as the idea of race did among the most hard-line psychometricians (Saini, 2019). All of these researchers maintained that intelligence was inherited and that it could be used to derive differences among populations, which could thus be treated as races.

In this history, Arthur R. Jensen (1923–2012), a kind of American Galtonian, played a significant role, becoming famous for his genetic and racist rearticulation of intelligence test differences between Blacks and Whites, especially regarding measures of general intelligence that were supposed to be culture-free. In 1969, Jensen published a lengthy study resurrecting the concept of race and linking it to the fact that African Americans did not achieve academic success and were poor; despite social assistance programs, their intelligence was argued to be genetically inferior, and the aid was considered a waste of public money (Jensen, 1969).

Jensen was the subject of condemnation by many colleagues and was harshly attacked even on a personal level, with calls for his dismissal. Nevertheless, he was also regarded as a master by psychometricians and received prizes and funding during his lifetime, particularly from the Pioneer Fund. This American association financed psychometric studies aimed at promoting segregationist and racist policies, which were active until recently and were managed by this group of psychometricians [the last president was Richard Lynn (1930–2023), a controversial English psychologist and eugenicist] (Tucker, 2007).

In Jensen's case, critics had an easy task because he used the data of Cyril Burt, another English psychometrician who, as previously noted, had fraudulently employed twin data on the heritability of intelligence. Jensen brought his own data alongside Burt's to legitimize his estimates of heritability. This was the sign that even in Jensen's work, the accounts did not add up; it seemed merely a sophistry to claim that intelligence was 80% genetic. Jensen, as well as Burt, received extensive methodological criticism.

The Cyril Burt affair marked a defeat for psychometricians, but they did not surrender. Eysenck himself, a student of Spearman and Burt, continued to collect data to demonstrate the inferiority of Blacks on tests, correlating it, for example, with neurophysiological indices to support the paradigm that intelligence was inherited mainly (the Pioneer Fund also funded Eysenck). In short, these studies perpetuated an Anglo-Saxon Nordicist vision (similar to that which had influenced Nazi eugenicists). Many criticisms could be leveled at these studies. Some examples include:

- They assumed the possible existence of an abstract intelligence as a general factor.

- Despite methodological efforts, the tests were never truly culture-free.
- They compared populations without accounting for within-population variability.
- They accepted correlations as causal.
- They always chose genetics as the primary cause determining intelligence, behaviors, culture, or even the backwardness of a population.
- They used indirect genetic indicators.[16]
- They disregarded environmental, economic, or cultural causes as determinants of the intelligence measures used.[17]

Moreover, these hard-line psychometricians were convinced that the tests were as precise as "thermometers" and not methodologically weak tools requiring critical use.

The science war was chiefly between these psychometricians and evolutionary biologists and cultural psychologists (Lewontin, Gould, Kamin), who criticized the naivety of the psychometricians' research in light of population genetics.[18] In my view, the war of the hard-line psychometricians was scientifically lost from the start because it was an indirect, Aristotelian-classificatory science, rather than a Galilean one, in the sense of Kurt Lewin (1890–1947). According to Lewin, in psychology as in physics, it was necessary to move from an abstract classificatory procedure (Aristotelian conception) to an essentially concrete, constructive Galilean method; reference to the totality of the situation had to replace the broad collection of frequent repetitions and abstract set construction.[19] This meant methodologically that the importance of a case and its validity as proof were not determined by its frequency of occurrence, as is the case

[16] A rather bizarre example is the use of penis length to demonstrate the existence of races (Lynn, 2013).

[17] For criticisms of Jensen and racist psychometricians, see the final chapter of Jackson and Weidman (2006).

[18] For the polemic between Eysenck and Kamin, see *Intelligence: The Battle for the Mind* (1981).

[19] For Lewin (1931), the Aristotelian approach failed to grasp the uniqueness of individual cases; it was based on frequency analysis and the inclusion of cases in sets. The Galilean approach excluded abstract classification of natural phenomena, proposing a methodical investigation of individual cases that led to a theory capable of explaining unique events and capturing dynamic psychic processes. Ultimately, Lewin criticized psychometrics for being static and incapable of offering actual scientific knowledge.

with research on samples. In short, the illusion of racist psychologists was the illusion of psychologists who unknowingly practiced an Aristotelian science that took correlations as causes, started from phenomena without questioning whether what our senses perceive might be an illusion, used self-referential prejudices as hypotheses to confirm, and constructed sets based on superficial elements that were artifacts later named in some way. Races were an example of such sets.

Despite these epistemic limits, the pseudoscience of race and intelligence colluded with the culture of neoliberal economics (Winston, 2018), whereby the state, in pursuit of investment, should always follow a (fallacious) zero-sum logic (zero-sum bias). Central to this logic is the belief that there is always a fixed sum to be distributed conventionally and that funding should be allocated where money is not "wasted," because giving to one person necessarily takes away from someone else. This type of logic appears self-evidently correct and is pervasive, becoming a kind of radical thought in which not only money, but also elements like rights, intelligence, and feelings, necessarily follow a zero-sum distribution (Różycka-Tran et al., 2015). If one group has more wealth, happiness, intelligence, or rights, then another group must necessarily lose wealth, happiness, intelligence, and rights. Implicitly assuming this belief, whenever something is ceded in a relationship, there is a pole that loses or perceives a loss (albeit fictitious). The fallacy of zero-sum thinking leads precisely to the belief in eugenics, according to which marriages between supposedly hierarchically different "races" should be restricted because mixing would lead to a psychological and intellectual deterioration of the more evolved populations rather than an actual improvement.

This becomes a problem because the values of democracy are based on promoting rights and happiness for all without imposing limits of zero-sum logic. Racist psychometricians gained notoriety precisely because their ideas seemed to justify those who followed such logic. If reality were immutable and intelligence naturally distributed unequally among populations considered "races," it made no sense to change a predetermined order made unalterable by nature itself. Attempting social program changes to a situation considered fixed further frightened those who feared their wealth or intelligence might be taken by those who had none.

Probably for these reasons, old correlations purporting to demonstrate heritability of intelligence continue to resurface cyclically (now with percentages lower than in Jensen's time). Above all, there is a persistent instrumental use of such data to delegitimize social investments and programs. If poor and racialized groups remain intellectually inferior and unchangeable, there is no need to invest in their improvement.

In 1994, *The Bell Curve* was published, a book by psychologist Richard Herrnstein and political scientist Charles Murray (Herrnstein & Murray, 1994). Their study aimed to demonstrate significant intelligence differences among social classes, as well as to reveal ethnic differences to prove, by this method, the heritability of IQ. They repeated Jensen's errors and poorly framed comparisons. Both Jensen's and The Bell Curve's studies were refuted by evidence controlling for socio-economic variables, such that IQ differences vanished. The causes of inequalities were poverty and the cultural background of the participants, rather than genetics (Cavalli-Sforza & Cavalli-Sforza, 2013, pp. 333–340). The Bell Curve also belonged to a group of studies explicitly designed to eliminate educational programs considered a waste of resources, as intelligence was viewed as an innate, limited, differentially distributed, and unchangeable phenomenon. The hidden objective of this study was to promote the idea that public money should not be widely distributed, but to avoid waste, financing should be reserved only for the best.

Eysenck, until his death, remained a respected psychologist but was convinced of Blacks' intellectual inferiority; more recently, Richard Lynn wrote numerous articles always attempting to prove that economic well-being totally depends on intelligence, considered a genetically determined factor, that southern populations (including in Italy) were poorer because less intelligent, and that their genetic heritage caused their lower intelligence (Lynn, 2013).

Moreover, the currently widespread evaluation culture often suffers from this pernicious Galtonian approach, whereby resources are allocated based on merit measures. In these evaluation studies, for example, school performance is treated similarly to intelligence tests; it becomes easy to find what one wants—that in poorer areas, children's performance is lower (Foschi, 2017).

In this culture of merit assessment (through testing), some international psychological articles, while criticizing Lynn, continued to publish unproven ideas that perpetuated the Nordicist cliché, for example that the Italian case should be studied from a genetic point of view, since the differences in intelligence among children in the north were positively derived from genetic mixing with central/northern Europeans, while the lower values in the south were due to mixing with North Africa (Cornoldi et al., 2013).[20] In short, ideas identical to those that could be read in Lombroso and other nineteenth-century researchers were still being published just a few years ago.

In the line of methodological critiques of these "discoveries" by psychometricians, James Flynn (1934–2020) highlighted one of the most interesting phenomena connected to intelligence: far from being a passive, static product of genetics, intelligence, as measured by tests, appears rather as a highly change-sensitive phenomenon. The Flynn effect, discovered in opposition to Jensen, involved the observation that over generations, the average IQ score of a population increased significantly and unexpectedly worldwide (by dozens of points in a few decades). Furthermore, in particular highly developed populations, the average IQ, after initially rising, has recently been decreasing unpredictably. Intelligence is thus a somewhat variable measure, and the cyclic dynamism of its variation appears to be attributable to historical-environmental factors such as nutrition, evolutionary well-being, schooling programs, cultural changes, and the type of questions used in tests, which can orient improvements or declines in performance from one generation to another. It should not be forgotten that in countries where average intelligence measures are lowest, in addition to an objective poverty problem, there are also serious infant infectious diseases impairing the normal psychophysical and neurological development of populations. In light of these very robust data, the evidence from hard-line psychometricians like Jensen becomes even more hollow in meaning.

Thus, pseudoscientific racism and prejudice seem like a metastatic cancer that lurks, seems defeated, but continually returns. In some cases, it is even hidden in things that seem most virtuous, such as merit-based evaluations, where the objective is to allocate resources to those assessed. This produces an eugenic mechanism by which those assessed worse

[20] Interestingly, despite fundamental criticism of Lynn's genetic hypothesis, the authors state: "the comparison between North and South Italy had the advantages of comparing two populations that were administered exactly the same tasks in the same language and with the same school legislation. The two populations are apparently different not only on a historical but also on a biological basis as South Italy is close to African communities and immigrations and North Italy has close exchanges with Central Europe [...] With the present paper we did not intend to question the general issue of heritability of intelligence, as, in our view, the evidence supporting the genetic bases of intelligence is robust and unquestionable [...]" (Cornoldi et al., 2013, p. 31).

receive less funding, and if they do not improve, are definitively disadvantaged. Evaluation programs that appear to ignore that funding the best increases rather than decreases differences. Such programs seem plausible only if their ultimate goal is to eliminate those with worse scores. The eliminationist logic we have already discussed returns, if only implicitly.

In recent years, following the continuing killings of African Americans by police, a strong movement has arisen that, in addition to removing statues dedicated to figures implicated in the history of slavery, has contributed to reviewing the names assigned to university chairs, spaces, and even lifetime achievement awards for scientists implicated in eugenics and racism. The hard-line psychometricians we have discussed have been specifically targeted. For example, while I was writing the first edition of this volume, Gregory L. Fenves, President of Emory University, based on a petition by professors and alumni of the same university, decided to rename the spaces and the chair honoring Robert Yerkes, one of the prominent experimental psychologists of the twentieth century, but who shared eugenic ideology.[21] Similarly, Lynn, whose research has been repeatedly criticized, had his emeritus professor title revoked before his death.[22]

In 2020, the Department of Psychology at the University of Western Ontario publicly released this statement regarding John P. Rushton (1943–2012), a former member of the same department, one of the leading psychometricians oriented toward racist perspectives, a friend of Jensen, Eysenck, Lynn, and the longtime director of the Pioneer Fund:

> In addition to ethical concerns about the nature and funding of his research, Rushton's work is deeply flawed from a scientific standpoint. Crucially, Rushton's works linking race and intelligence are based on an incorrect assumption that fuels systemic racism, the notion that racialized groups are concordant with patterns of human ancestry and genetic population structure. This idea is rejected by analysis of the human genome: racialized groups are not distinct genetic populations (Yudell et al., 2016). What Rushton described as "races" are socially created categories that do

[21] The petition concerned not only Yerkes but also L.Q.C. Lamar, a racist jurist and Democratic Party politician. One of the petition letters to the University Rector published in the university newspaper can be read here: https://emorywheel.com/letter-to-the-editor-concerning-president-fenvess-failure-to-respond-to-calls-to-remove-emorys-tributes-to-l-q-c-lamar-and-robert-yerkes/

[22] https://www.bbc.com/news/uk-northern-ireland-43768132.

not reflect patterns of human inheritance or genetic population structure. In an effort to explain putative differences in parental care between racialized groups, Rushton also inappropriately applied an ecological theory developed to explain differences between species' reproductive strategies (r/K selection theory), an approach that has since been thoroughly debunked (Allen et al., 1992; Anderson, 1991). Moreover, Rushton's work is characterized by a complete misunderstanding of population genetic measures, including fundamental misconceptions about the nature of heritability and gene-environment interactions during development. His work has been criticized, often by other Western University faculty members (Bailey, 1997; Cain & Vanderwolf, 1990), on many other grounds. In some cases, Rushton's work has failed to replicate or withstand reanalysis (Cernovsky & Litman, 1993; Peregrine et al., 2003). In other cases, his papers ignored alternative explanations or competing evidence that did not support his racist hypotheses. (Cain & Vanderwolf, 1990)
Despite its deeply flawed assumptions and methodologies, Rushton's work and other so-called "race science" (currently under the pseudonym of "race realism") continue to be misused by white supremacists and promoted by eugenic organizations. Thus, Rushton's legacy shows that the impact of flawed science lingers on, even after qualified scholars have condemned its scientific integrity. Academic freedom and freedom of expression are critical to free scientific inquiry. However, the notion of academic freedom is disrespected and abused when it is used to promote the dissemination of racist and discriminatory concepts. Scientists have an obligation to society to speak loudly and actively in opposition of such abuse.[23]

References

Adorno, T. W., Frenkel-Brunswik, E., Levinson, D. J., & Sanford, R. N. (1950). *The authoritarian personality.* Harpers.

Allegri, G., & Ciccarelli, R. (2013). *Il quinto stato* [The fifth estate]. Ponte alle Grazie.

Allen, G., Eriksson, A. W., Fellman, J., Parisi, P., & Vandenberg, S. G. (1992). Twinning and the r/K reproductive strategy: A critique of Rushton's theory. *Acta Geneticae Medicae et Gemellologiae: Twin Research, 41*(1), 73–83.

American Psychiatric Association. (1974). *Diagnostic and statistical manual of mental disorders(DSM-II)* (2nd ed.). American Psychiatric Association.

[23] https://www.psychology.uwo.ca/people/faculty/remembrance/rushton.html.

Anderson, J. L. (1991). Rushton's racial comparisons: An ecological critique of theory and method. *Canadian Psychology, 32*, 73–83.

Anderson, J., & Franklin, B. (1734). *The constitution of the free-masons* (Philadelphia ed.). UNL Libraries: Faculty Publications.

Bailey, R. C. (1997). Hereditarian scientific fallacies. *Genetica, 99*(2–3), 125–133.

Cain, D. P., & Vanderwolf, C. H. (1990). A critique of Rushton on race, brain size and intelligence. *Personality and Individual Differences, 11*(8), 777–784.

Cavalli-Sforza, L. L., & Cavalli-Sforza, F. (2013). *Chi siamo: la storia della diversità umana* [Who we are: The history of human diversity]. Codice edizioni.

Cernovsky, Z. Z., & Litman, L. C. (1993). Re-analyses of J. P. Rushton's crime data. *Canadian Journal of Criminology, 35*(1), 31–36.

Cicciola, E. (2019). *La scoperta dell'intelligenza. Alfred Binet e la storia del primo test* [The discovery of intelligence: Alfred Binet and the story of the first test]. Fefè.

Clark, K., & Clark, M. (1950). The Negro child in the American social order. *The Journal of Negro Education, 19*, 341–350.

Clemente, P. (1971). *Frantz Fanon fra esistenzialismo e rivoluzione* [Frantz Fanon between existentialism and revolution]. Laterza.

Cornoldi, C., Giofrè, D., & Martini, A. (2013). Problems in deriving Italian regional differences in intelligence from 2009 PISA data. *Intelligence, 41*(1), 25–33.

De Leo, M. (2021). *Queer. Storia culturale della comunità LGBT+* [Queer. A cultural history of the LGBT+ community]. Einaudi.

Du Bois, W. E. B. (1939). *Black folk: Then and now: An essay in the history and sociology of the Negro race*. Holt, Rinehart & Winstone.

Evola, J. (1968–2000). *L'arco e la clava* [The bow and the club]. Mediterranee.

Evola, J. (1969). Avvento del «Quinto Stato»? [Advent of the "Fifth Estate"?]. *Il Borghese, XX*(30), 684–689.

Evola, J. (1997). *Metafisica del sesso* [Metaphysics of sex]. Mediterranee.

Eysenck, H. J., & Kamin, L. (1981). *Intelligence: The battle for the mind*. Macmillan.

Faderman, L. (2015). *The gay revolution: The story of the struggle*. Simon & Schuster.

Fanon, F. (1961). *Les damnés de la terre* [The wretched of the earth]. Maspero.

Foschi, R. (2017). Vino vecchio in botti nuove: I test di intelligenza, la valutazione scolastica e il benessere della società [Old wine in new barrels: Intelligence tests, school assessment, and social well-being]. *International Journal of Psychoanalysis and Education, 9*(1), 47–54.

Foschi, R. (2024). "He who leaves the old way for the new, knows what he leaves, but does not know what he will find": Traditionalism, zero-sum thinking, and imagined future. *World Futures, 80*(3), 223–240.

Franklin, B. (1763, December 17). *From Benjamin Franklin to John Waring*. Founders Online, National Archives. https://founders.archives.gov/documents/Franklin/01-10-02-0214

Fryer, J. (1985). The unveiling of Dr. Anonymous. *Association of Gay and Lesbian Psychiatrists, 11*(2), 1–4.

Hayden, L. (1866). *Caste among Masons: Address before Prince Hall Grand Lodge of free and accepted Masons of the State of Massachusetts, at the Festival of St. John the Evangelist, December 27, 1865*. Coombs & Co.

Herrnstein, R. J., & Murray, C. (1994). *The bell curve: Intelligence and class structure in American life*. Free Press.

Inglehart, R. (1977). *The silent revolution: Changing values and political styles among Western publics*. Princeton University Press.

Inglehart, R. F., & Norris, P. (2016). *Trump, Brexit, and the rise of populism: Economic have-nots and cultural backlash*. Harvard University, John F. Kennedy School of Government.

Jackson, J. P., & Weidman, N. M. (2006). *Race, racism, and science: Social impact and interaction*. Rutgers University Press.

Jefferson, T. (1787). *Notes on the State of Virginia*. Stockdale.

Jensen, A. (1969). How much can we boost I.Q. and scholastic achievement? *Harvard Educational Review, 39*(1), 1–123.

Jost, J. (2020). *A theory of system justification*. Harvard University Press.

Kantrowitz, S. (2010). 'Intended for the better government of man': The political history of African American Freemasonry in the era of emancipation. *The Journal of American History, 96*(4), 1001–1026.

Kuznick, P. J. (2019). *Beyond the laboratory: Scientists as political activists in 1930s America*. University of Chicago Press.

Lachman, G. (2018). *Dark star rising: Magick and power in the age of Trump*. TarcherPerigree.

Lauriola, M., Foschi, R., & Marchegiani, L. (2015). Integrating values and cognitive style in a model of right-wing radicalism. *Personality and Individual Differences, 75*, 147–153.

Lewin, K. (1931). The conflict between Aristotelian and Galilean modes of thought in contemporary psychology. *Journal of Genetic Psychology, 5*, 141–177.

Liss, J. E. (1998). Diasporic identities: The science and politics of race in the work of Franz Boas and W. E. B. Du Bois, 1894–1919. *Cultural Anthropology, 13*(2), 127–166.

Lynn, R. (2013). Rushton's r–K life history theory of race differences in penis length and circumference examined in 113 populations. *Personality and Individual Differences, 55*(3), 261–266.

Milgram, S. (1974). *Obedience to authority: An experimental view*. Harper & Row.

Montagu, A. (1952). *Man's most dangerous myth. The fallacy of race. Third edition—Revised and enlarged*. Harper & Brothers.

Montuori, A. (2005). How to make enemies and influence people: Anatomy of the anti-pluralist, totalitarian mindset. *Futures, 37*(1), 18–38.

Moretti, E. (2021). *The best weapon for peace: Maria Montessori, education, and children's rights*. University of Wisconsin Press.

Myrdal, G. (1944). *An American dilemma: The negro problem and modern democracy*. Harper & Bros.

Nash, G. B. (2006). Franklin and slavery. *Proceedings of the American Philosophical Society, 150*(4), 618–635.

Norris, P. & Inglehart, R. (2019). *Cultural Backlash: Trump, Brexit and Authoritarian Populism*. Cambridge University Press.

Różycka-Tran, B., Boski, P., & Wojciszke, B. (2015). Belief in a zero-sum game as a social axiom: A 37-nation study. *Journal of Cross-Cultural Psychology, 46*(4), 525–554.

Saini, A. (2019). *Superior: The return of race science*. Beacon Press.

Spektorowski, A., & Mizrachi, E. (2004). Eugenics and the welfare state in Sweden: The politics of social margins and the idea of a productive society. *Journal of Contemporary History, 39*(3), 333–352.

Stanton, E. C. (1889). *A history of woman suffrage* (vol. 1). Fowler & Wells.

Tucker, W. H. (2007). *The funding of scientific racism: Wickliffe Draper and the pioneer fund*. University of Illinois Press.

UNESCO. (1950). *The race question*. UNESCO Publication 791. https://unesdoc.unesco.org/ark:/48223/pf0000128291

UNESCO. (1969). *Four statements on the race question*. UNESCO. https://unesdoc.unesco.org/ark:/48223/pf0000122962

UNESCO. (1979). *Declaration on race and racial prejudice adopted by the General Conference of Unesco at its twentieth session Paris, 27 November 1978*. UNESCO. https://unesdoc.unesco.org/ark:/48223/pf0000039429

Winston, A. S. (2018). Neoliberalism and IQ: Naturalizing economic and racial inequality. *Theory & Psychology, 28*(5), 600–618.

Yudell, M., Roberts, D., DeSalle, R., & Tishkoff, S. (2016). Taking race out of human genetics. *Science, 351*(6273), 564–565.

CHAPTER 9

The Heartbreaking and Wonderful Beauty

Geneticists and biologists have provided the most compelling explanations for why the concept of race does not exist as a biological reality in *Homo sapiens*. One of the most important figures was Richard C. Lewontin (1929–2021), who first highlighted a crucial argument regarding the unreality of human races. By studying specific known genes and using as samples those groups which, at the time, anthropologists tended to consider different human populations (the traditional races), he discovered that about 85% of genetic variability lay "within" what were considered racial groups, with the remaining variability, around 15%, relating to differences between the alleged races and subgroups:

> It is clear that our perception of relatively large differences between human races and subgroups, as compared to the variation within these groups, is indeed a biased perception and that, based on randomly chosen genetic differences, human races and populations are remarkably similar to each other, with the largest part by far of human variation being accounted for by the differences between individuals.
>
> Human racial classification is of no social value and is positively destructive of social and human relations. Since such racial classification is now seen to be of virtually no genetic or taxonomic significance either, no justification can be offered for its continuance. (Lewontin, 1972, p. 397)

Races thus appear different to us due to purely phenotypic and superficial elements (such as skin color and facial shapes). However, these phenotypic

R. Foschi, *Racisms*, https://doi.org/10.1007/978-3-032-22962-5_9

differences do not correspond equivalently in the genome. If races existed, they would have to be internally similar yet distinct from one another. Genetic variability should be greater between groups than within groups. The data have been verified repeatedly with more direct techniques, and the results remain similar and robust. The famous 15% variability between different groups also appears unusable in itself for proposing geographical or phenotypic races (Barbujani, 2005, 2006). Using the most up-to-date population genetics methods, the old racial distinctions have thus fallen away.

Genetic differences between ethnicities that have been more separated are also minimal, and it is hypothesized that if mixing between Whites and Blacks were favored, skin color would become uniform within a few generations. For all these reasons, the physical-biological aspects of humans can be placed on a continuous rather than a discrete scale (Cavalli-Sforza & Cavalli-Sforza, 2013).

Although the existence of human races is not empirically supportable,[1] racisms and prejudices against human groups remain a very present reality. Moreover, this volume could not cover the history of all racial or similar prejudices, which is truly impressive. At every moment in history when a scapegoat was needed to justify a more or less catastrophic event (such as Germany's defeat in World War I), racism was the preferred ideology to "project" the causes of adversity outside one's own responsibility.

Racisms are also a ubiquitous phenomenon in all cultures and continents. The psychological polarization between "us-good" and "them-bad" forms the basis for the construction of these scapegoats. Europeans and Americans have been among the most racist populations in history. Colonialism was the policy through which entire peoples, such as Native Americans, were subjugated and destroyed, or Africans enslaved. Japan was another culture that considered itself hegemonic vis-à-vis other Eastern peoples; in Africa itself, racial hatred between Tutsis and Hutus produced some hundreds of thousands of deaths in a fratricidal war that erupted in the 1990s. In India, the caste system appears to be an immutable fact accepted even by the most educated population. In Italy, there are prejudices between the North and South that have given rise to

[1] Saini (2019) reports that after asking Luca Cavalli-Sforza (1922–2018) whether he considered the category of race valid, he received the blunt email reply: "There are simply no races in humankind" (p. 113).

political ideologies justifying the economic crisis. Sexist and homophobic discrimination continues.

Examples could be numerous, and a racist worldview can also be connected to contemporary populisms, considered an identity syndrome (Genovese, 2021). When a group such as Whites loses power over other ethnic groups, men over women, heterosexuals over homosexuals, this sort of identity pathology emerges, founded on grievance by the group perceiving itself as dispossessed of power. This can in turn lead to racisms. At best, this grievance halts at the formation of populist political movements opposing those who would finance and foster power shifts experienced as threats and degenerations of democracy; at worst, racism arms the hands of young white supremacists who randomly kill persons ethnically different from themselves. This phenomenon, alongside racial discrimination, is particularly present in the United States, for example, in violent police actions.

As we have attempted to demonstrate throughout this treatment, racisms manifest with extreme variability; perhaps there exists a sort of "racist spectrum" that ranges from paternalistic prejudices to hatred, passing through populist demands. Prejudices, which are also being fought today, however, with a universalistic vocation typical of contemporary times. A clear example is the guiding principles of the Black Lives Matter movement, born in response to police violence against African Americans.[2] While starting from the classic themes of the civil rights movement, it addresses a broad range of issues related to combating racisms and prejudices:

Unapologetically Black

We are unapologetically Black in our positioning. In affirming that Black Lives Matter, we need not qualify our position. To love and desire freedom and justice for ourselves is a prerequisite for wanting the same for others.

[2] The Black Lives Matter movement/project was created in 2013 by three African American women: Alicia Garza, Patrisse Cullors, and Opal Tometi.

All Black Lives Matter

We are guided by the fact that all Black lives matter, regardless of actual or perceived sexual identity, gender identity, gender expression, economic status, ability, disability, religious beliefs or disbeliefs, immigration status, or location.

Globalism

We see ourselves as part of the global Black family, and we are aware of the different ways we are impacted or privileged as Black people who live in different parts of the world.

Diversity

We acknowledge, respect, and celebrate differences and commonalities. We work vigorously for freedom and justice for Black people and, by extension, all people.

Restorative Justice and Struggle

We intentionally build and nurture a beloved community that is bonded together through a beautiful struggle that is restorative, not depleting.

Trans Affirmation

We make space for transgender siblings to participate and lead. We are self-reflexive and do the work required to dismantle cisgender privilege and uplift Black trans folk, especially Black trans women who continue to be disproportionately impacted by trans-antagonistic violence [...] (https://www.blmla.org/guiding-principles).

Paradoxically, civil rights struggles were also based on an us-versus-them polarization, Black versus White, poor versus rich. In these struggles, marginalized people regained power by identifying with the group they were classified in. The perception that races were biological entities existed on both sides of the conflict. These anti-racist struggles and their subsequent identity-based reactions remind us that even in liberal and democratic societies, prejudices and hierarchical conceptions rewarding those who decide the rules of the game are still very present. Although in the most advanced constitutional charters, such as the Italian one, any possibility of tolerance toward discrimination and racism is denied (for example, the Italian Constitution in Article 3, paragraph 1, affirms that "All citizens have equal social dignity and are equal before the law without distinction of sex, race, language, religion, political opinions, personal and social conditions"), experience and history teach that this presumption is often disregarded in practice.

When new social groups emerge with whom to share one's existence, traditionally hegemonic groups can transgress the foundational principles of democracies. The newcomers are perceived as a danger to society and are to be stopped by immigration regulations, segregationist attitudes, defunding of educational and anti-poverty programs, and evaluations that reward those who already have, adhering to the logic of "those who have not, are not."[3] The idea is that wherever real or perceived threats arise, specific racist beliefs and discriminatory behaviors may be triggered. Individuals tend to justify discrimination against other groups by asserting that they are adversaries, by appropriating resources or enjoying rights perceived as undeserved (as is typically the case, for instance, regarding immigrants). The perceived threat is a significant and transversal variable, and one can find cognitive antecedents in social axioms such as the already cited zero-sum fallacy. As we have already seen, this conception—that since resources are limited, any material or psychological reality becomes a zero-sum game in which one's prosperity necessarily means deprivation for another—renders it automatic for an individual with such beliefs to attribute the cause of their deprivation to a third party, different from themselves, who is appropriating resources that should belong to them. Thus, a real event or situation, leveraging preexisting prejudices, activates, in a competitive situation, what one might define as a *trigger*. This psychological (cognitive and emotional) mechanism initiates the establishment of racist or populist discourse (Vincenzo & Foschi, 2021).

As we saw in the previous chapter, zero-sum logic can also be applied to the rigid theory of the normal distribution of intelligence. So, suppose conservatives perceive the situation they live in as a defeat. In that case, those responsible for this defeat are identified as a group of people who, according to zero-sum thinking, take something away from them (wealth, power, happiness, sexual power, etc.) and are perceived as different from the hegemonic group and easily racialized. These groups, perceived as potentially "revolutionary," can thus be dehumanized and treated on a hierarchical scale as inferior and responsible for changing the status quo. Feeling cheated out of important parts of one's identity thus leads to

[3] "Chi non ha, non è" is a Calabrian aphorism made famous by Franco Basaglia (1924–1980) in an interview with Sergio Zavoli (1923–2020), later featured in the Italian television show, *I giardini di Abele* (The gardens of Abel) (1968).

the search for scapegoats, identified with certain classes of people, which may vary as social needs change. Nostalgia for the hierarchies of the past, on the other hand, provides reassurance in the face of social developments that have led to the emancipation of workers, enslaved people, women, immigrants, and homosexuals. This seems to be the psychological mechanism behind the backlash of right-wing populism (Inglehart & Norris, 2016), in which "conservatives tend to view the gains of some groups (e.g., women, African Americans, and immigrants) as offset by other groups' losses (e.g., men, European-Americans, and U.S. citizens). Conservatives, for example, are more likely to believe that expanding civil rights for minorities comes at the expense of the majority and that increasing job opportunities for women diminishes opportunities for men" (Davidai & Ongis, 2019, p. 1; see also Roberts & Davidai, 2022; Davidai & Tepper, 2023).

This sense of loss is surprisingly similar to that which develops in the Oedipus complex on the part of the father who feels usurped by the newborn. In the opening minutes of his film Oedipus Rex, Pier Paolo Pasolini (1922–1975), in order to express the father's feelings toward his son, inserts a black inscription that reads: "You are here to take my place in the world, to push me back into nothingness and steal everything I have." In this sense, the logic of Oedipus has a more general value than the more specific and familiar meaning attributed to it by Freud, and is very similar to zero-sum thinking.

Nature, however, does not operate according to zero-sum logic; it does not treat the allocation of resources as potential waste. Society would likely function better if it avoided the distortions created by zero-sum thinking. In nature, extreme variability and vast potential are evident, and biodiversity is its defining feature. The very mechanisms illuminated by Darwin place variability at the foundation of evolution. Attempting to steer or tame the nature of variability tends not to respect how nature actually functions. In this sense, biodiversity is the precise opposite of eugenics, which naïvely sought to control evolution according to a zero-sum fallacy—that is, to eliminate the genetic heritage of certain 'undesirable' human groups while correspondingly increasing that of "desired" groups, as if the heritage of some necessarily diminished the heritage of others. Today, genetics demonstrates an extreme complexity in the potential of our genetic heritage, which continuously hybridizes through sexual reproduction and cultural interactions, so that diversities increasingly enrich one another. The world is changing rapidly, and humans appear to have the potential to drive these changes. Paradoxically, attempting to domesticate the process—for example, by controlling mating and reducing genetic variability—could eliminate latent traits that might protect us from certain diseases, potentially worsening rather than improving the situation:

> We need to maintain existing genetic diversity because we cannot predict the challenges that may await us in the future, particularly new infectious diseases. Terrible germs may always appear on the scene, as with the HIV virus. We know little about possible individual variability in predisposition to this disease; however, such variability is typically present for all infectious diseases. If we suppressed these variations and, by chance, bred individuals sensitive to a new severe disease, it could mean the end of humanity.
> By arbitrarily selecting some type that seems particularly promising, we can profoundly alter the species. (Cavalli-Sforza & Cavalli-Sforza, 2013, p. 393)

A eugenic plan, therefore, always carries the risk of being both ineffective and dangerous. In this context, the scientific reasoning applied by humans often seeks to "tame" nature's complex probabilistic laws (Hacking, 1990), assuming that the biological and psychological nature of individuals follows an entirely predictable logic. In doing so, humans place nature itself in a kind of Procrustean bed, leading to oversimplifications and rigid schematizations. While simplification may seem helpful in managing complexity and controlling new social phenomena—such as the emergence of historically marginalized groups claiming their rights—it can ultimately divert us from the proper management of complex social and political dynamics, with potentially serious consequences. Understanding complex systems is far from trivial.

Races, as biologists explain, were one of these schematizations, and they do not exist because humans are genetically similar while also exhibiting considerable individual diversity. In any case, humans cannot be distinguished by discrete genetic criteria as is possible in some animal species. The belief in races has paradoxically constrained both equality and diversity. Humans, all belonging to a single species, share basic biological and psychological systems, which then manifest at both individual and group levels in complex and diverse ways not directly determined by genetics. Furthermore, humans demonstrate enormous potential, capable of unexpected adaptations and significant evolutionary changes; indeed, genetic heritage still holds many surprises and modes of interaction with the environment that we are only beginning to understand.[4] Therefore, the use still made of the notion of race, for example, in statistical and medical research to describe a sample of individuals susceptible to a disease or a therapy (Kanakamedala & Haga, 2012), is misleading. As history has shown, it can lend itself to racist conceptions and further discrimination, besides distancing us from the understanding of reality.[5]

As complexity theorists have shown, when faced with dynamic phenomena, we often resort to defensive mechanisms that ineffectively attempt to constrain them. This is especially true regarding the diversity of human nature, so that the other, those different from us, trigger phenomena similar to those analyzed by Else Frenkel-Brunswik and Adorno's group. Faced with populations or cultures that seem out of tune with our traditions, people can become intolerant at all levels, and in extreme cases, history has shown that humans can resort to inhumane political and pseudoscientific solutions. Intolerance of ambiguity drives toward schematization; the reduction of complexity and anxiety provoked by the perception of what seems ambiguous or chaotic may lead to the formation of a pervasive, polarizing, and totalitarian mentality (Montuori, 2005).

Education for tolerance, discussion of phenomena according to the logics of complexity, and the search for creative solutions to deconstruct racisms and prejudices must not be considered ideological options. Our discussion shows that these solutions arise from what history and scientific research teach us about the problem of racism.

In the remarkable mid-length film *What Are Clouds?* (Che cosa sono le nuvole?) by Pier Paolo Pasolini,[6] the tragedy of Othello is reenacted with actors serving as puppets. The actors Ninetto Davoli and Totò—respectively the Moor, whose puppet has a black-painted face, and Iago, whose face is green—embody human complexity against a backdrop of racial and gender issues. In an effort to overturn racist clichés, Pasolini inverts conventional stereotypes: the Moor is portrayed as virtuous and

[4] See contemporary epigenetics studies. On these topics, see also Rose (2007).

[5] Saini recently also described the problems related to contemporary attempts to racialize pharmacological research, which would fail to yield expected results (2019).

[6] This refers to an episode within the anthology film *Capriccio all'Italiana* (Caprice Italian Style) (1968).

deceived.[7] At one point, Othello asks Iago why he is widely hated and why he has deceived him. Shakespeare's text offers no answer. In his final appearance in the play, Iago refuses any explanation: "Demand me nothing: what you know, you know. From this time forth I never will speak word."

Pasolini, however, radically alters the conclusion of the tragedy. He imagines the two protagonists as worn-out puppets who have never seen the sky. In the final scene, the author consigns both of them to an open-air landfill, where they lie together under a bright sky. Astonished, Othello asks Iago, "And what are those?" Iago replies, "Those are clouds." "And what are clouds?" Othello persists. Iago can only answer, "I do not know." "How beautiful they are! How beautiful they are!" exclaims a smiling Othello. The drama ends with Iago's final, unexpected outcry: "Ah! The heartbreaking and wonderful beauty of creation."

In the end, Pasolini suggests that nature, in all its diversity, possesses a beauty so overwhelming that it becomes at once disorienting and "heartbreaking." The filmmaker captures the condition of estrangement experienced by human beings when confronted with the wonders and multiplicity of the natural world and of human nature itself. This complexity elicits varied responses and remains profoundly challenging to comprehend fully.

In summary, racism today seems to be a reaction triggered by issues of identity. The sense of loss of power and identity leads people to identify diversity as the cause of that loss. These triggers can be set off by natural human biodiversity, which cannot be described as the presence of human races, but rather as the complexity of humanity's psychological and physical nature. This complexity easily stimulates a reaction that triggers conflicts, which (today) it is believed can be resolved with populist and racist narratives and justifications, hypothesizing a return to an idealized past, fidelity to tradition, the restoration of supposed greatness, and recourse to reassuring concepts in the face of a contemporary reality that presents complex crises.

[7] Critics immediately noted the predominance of race issues in Shakespeare's Othello; Desdemona and Othello have a secret relationship because they violate the prohibition against uniting two different "races." From this exceptional event, Iago's revenge unfolds, leading Othello to kill Desdemona, believing her unfaithful. Many have written on these themes; see, e.g., Kiernan (2016).

Overcoming the hidden logic of racisms, therefore, requires historical awareness but also an understanding of the complexity inherent in contemporary life, a process of understanding that is useful for creating a society in which diversity and pluralism are valued rather than perceived as threats.

References

Barbujani, G. (2005). Human races: Classifying people vs. understanding diversity. *Current Genomics, 6*, 215–226.
Barbujani, G. (2006). *L'invenzione delle razze: Capire la biodiversità umana* [The invention of races: Understanding human biodiversity]. Bompiani.
Cavalli-Sforza, L. L., & Cavalli-Sforza, F. (2013). *Chi siamo: la storia della diversità umana* [Who we are: The history of human diversity]. Codice edizioni.
Davidai, S., & Ongis, M. (2019). The politics of zero-sum thinking: The relationship between political ideology and the belief that life is a zero-sum game. *Science Advances, 5*(12), eaay3761.
Davidai, S., & Tepper, S. J. (2023). The psychology of zero-sum beliefs. *Nature Reviews Psychology, 2*(8), 472–482.
Genovese, R. (Ed.). (2021). *Sulla sindrome identitaria* [On the identity syndrome]. Rosenberg & Sellier.
Hacking, I. (1990). *The taming of chance*. Cambridge University Press.
Inglehart, R. F., & Norris, P. (2016). *Trump, Brexit, and the rise of populism: Economic have-nots and cultural backlash*. Harvard University, John F. Kennedy School of Government.
Kanakamedala, P., & Haga, S. B. (2012). Characterization of clinical study populations by race and ethnicity in biomedical literature. *Ethnicity & Disease, 22*(1), 96–101.
Kiernan, R. (2016). Racism, misogyny and 'motiveless malignity' in Othello. In *Discovering Literature: Shakespeare & renaissance*. British Library.
Lewontin, R. C. (1972). The apportionment of human diversity. In T. Dobzhansky, M. K. Hecht, & W. C. Steere (Eds.), *Evolutionary biology* (pp. 381–398). Springer.
Montuori, A. (2005). How to make enemies and influence people: Anatomy of the anti-pluralist, totalitarian mindset. *Futures, 37*(1), 18–38.
Roberts, R., & Davidai, S. (2022). The psychology of asymmetric zero-sum beliefs. *Journal of Personality and Social Psychology, 123*(3), 559–575.
Rose, N. (2007). *The politics of life itself: Biomedicine, power, and subjectivity in the twenty-first century*. Princeton University Press.
Saini, A. (2019). *Superior: The return of race science*. Beacon Press.

Vincenzo, C., & Foschi, R. (2021). *Razzismo vecchio stile, razzismo moderno, populismo: una ricerca di psicologia sociale* [Old style racism, modern racism, populism: A social psychology study]. In R. Genovese (Ed.), *Sulla sindrome identitaria* [On the identity syndrome] (pp. 23–70). Rosenberg & Sellier.

References

Addeo, F. (2008). *È normale la curva normale?* [Is the normal curve normal?]. Bonanno.

Adorno, T. W., Frenkel-Brunswik, E., Levinson, D. J., & Sanford, R. N. (1950). *The authoritarian personality.* Harpers.

Allegri, G., & Ciccarelli, R. (2013). *Il quinto stato* [The fifth estate]. Ponte alle Grazie.

Allen, G., Eriksson, A. W., Fellman, J., Parisi, P., & Vandenberg, S. G. (1992). Twinning and the r/K reproductive strategy: A critique of Rushton's theory. *Acta Geneticae Medicae Et Gemellologiae: Twin Research, 41*(1), 73–83.

American Eugenics Society. (1926). *A eugenics catechism.* American Eugenics Society.

American Psychiatric Association. (1974). *Diagnostic and statistical manual of mental disorders, second edition (DSM-II).* American Psychiatric Association.

Anderson, J. L. (1991). Rushton's racial comparisons: An ecological critique of theory and method. *Canadian Psychology, 32*, 73–83.

Anderson, J., & Franklin, B. (1734). *The constitution of the free-masons.* Philadelphia edition. UNL Libraries: Faculty Publications.

Azara, L., & Tedesco, L. (Eds.). (2020). *La donna delinquente e la prostituta: L'eredità di Lombroso nella cultura e nella società italiane* [The female delinquent and the prostitute: Lombroso's legacy in Italian culture and society]. Viella Libreria Editrice.

Babini, V. P. (1999). Un altro genere. La costruzione scientifica della «natura femminile» [Another gender: The scientific construction of "female nature"]. In A. Burgio (Ed.), *Nel nome della razza. Il razzismo nella storia d'Italia, 1870–1945* [In the name of race: Racism in Italian history, 1870–1945] (pp. 475–489). Il Mulino.

R. Foschi, *Racisms*, https://doi.org/10.1007/978-3-032-22962-5

Babini, V. P., Minuz, F., & Tagliavini, A. (1986). *La donna nelle scienze dell'uomo* [Woman in the human sciences]. FrancoAngeli.

Bailey, R. C. (1997). Hereditarian scientific fallacies. *Genetica, 99*(2–3), 125–133.

Banaji, M. R., & Greenwald, A. G. (2013). *Blindspot: Hidden biases of good people*. Delacorte Press.

Banfield, E. C. (1958). *The moral basis of a backward society*. Free Press.

Barbujani, G. (2005). Human races: Classifying people vs. understanding diversity. *Current Genomics, 6*, 215–226.

Barbujani, G. (2006). *L'invenzione delle razze: Capire la biodiversità umana* [The invention of races: Understanding human biodiversity]. Bompiani.

Barbujani, G. (2016). *Gli africani siamo noi: alle origini dell'uomo* [We are the Africans: At the origins of humankind]. Laterza.

Barbujani, G., & Cheli, P. (2022). *Sono Razzista ma sto cercando di smettere* [I'm racist but I'm trying to stop]. Laterza.

Baron-Cohen, S. (2018). Hans Asperger's Nazi collusion. *Nature, 557*, 305–306.

Battini, M. (2010). *Il socialismo degli imbecilli. Propaganda, falsificazione, persecuzione degli ebrei* [The socialism of fools: Propaganda, falsification, and persecution of Jews]. Bollati Boringhieri.

Bernays, E. L. (1928). *Propaganda*. Liveright.

Blumenbach, J. F. (1775). *De generis humani varietate nativa* [On the natural variety of the human species]. apud Vandenhoek et Ruprecht.

Blumenbach, J. F. (1865). *The anthropological treatises of Johann Friedrich Blumenbach* (Thomas Bendyshe, Trans. and Ed.). Anthropological Society. (Original work published 1795).

Brigham, C. C. (1923). *A study of American intelligence*. Princeton University Press.

Broberg, G., & Roll-Hansen, N. (1996). *Eugenics and the welfare state: Sterilization policy in Denmark, Sweden, Norway, and Finland*. Michigan State University Press.

Broca, P. (1861). Sur le volume et la forme du cerveau suivant les individus et suivant les races [On the volume and shape of the brain according to individuals and races]. *Bulletin de la Société d'Anthropologie* (Paris), *2*, 139–204, 301–321, 441–446.

Burgio, A., & Gabrielli, G. (2012). *Il razzismo* [Racism]. Ediesse.

Cain, D. P., & Vanderwolf, C. H. (1990). A critique of Rushton on race, brain size and intelligence. *Personality and Individual Differences, 11*(8), 777–784.

Carson, J. (2007). *The measure of merit: Talents, intelligence, and inequality in the French and American Republics, 1750–1940*. Princeton University Press.

Cassata, F. (2003). *A destra del fascismo. Profilo politico di Julius Evola* [To the right of fascism: Political profile of Julius Evola]. Boringhieri.

Cavalli-Sforza, L. L., & Cavalli-Sforza, F. (2013). *Chi siamo: la storia della diversità umana* [Who we are: The history of human diversity]. Codice edizioni.

Chamberlain, H. S. (1899). *Die Grundlagen des Neunzehnten Jahrhunderts* [The foundations of the nineteenth century]. Bruckmann.

Chase, A. (1977). *The legacy of Malthus: The social costs of the new scientific racism.* Knopf.

Cicciola, E. (2019). *La scoperta dell'intelligenza. Alfred Binet e la storia del primo test* [The discovery of intelligence: Alfred Binet and the story of the first test]. Fefè.

Cicciola, E., & Foschi, R. (2017). Giuseppe Sergi tra pensiero positivista e impegno politico [Giuseppe Sergi between positivist thought and political engagement]. *Physis, LII*, 169–182.

Ciracì, F. (2024). *Sul razzismo. Strutture logiche e paradigmi storico-filosofici* [On racism. Logical structures and historical-philosophical paradigms]. Mimesis.

Clark, K., & Clark, M. (1950). The Negro child in the American social order. *The Journal of Negro Education, 19*, 341–350.

Clemente, P. (1971). *Frantz Fanon fra esistenzialismo e rivoluzione* [Frantz Fanon between existentialism and revolution]. Laterza.

Colajanni, N. (1894). *La delinquenza in Sicilia e le sue cause* [Delinquency in Sicily and its causes]. Sandron.

Colajanni, N. (1898). *Per la razza maledetta* [For the accursed race]. Sandron.

Colajanni, N. (1903). *Razze inferiori e razze superiori o Latini e Anglo-Sassoni* [Inferior races and superior races or Latins and Anglo-Saxons]. Rivista Popolare.

Colla, P. S. (2000). *Per la nazione e per la razza: cittadini ed esclusi nel modello svedese* [For the nation and for the race: Citizens and excluded in the Swedish model]. Carocci.

Cornoldi, C., Giofrè, D., & Martini, A. (2013). Problems in deriving Italian regional differences in intelligence from 2009 PISA data. *Intelligence, 41*(1), 25–33.

Crenshaw, K. (1989). Demarginalizing the intersection of race and sex: A Black feminist critique of antidiscrimination doctrine, feminist theory and antiracist politics. *University of Chicago Legal Forum, 140*(1), 139–167. Available at: http://chicagounbound.uchicago.edu/uclf/vol1989/iss1/8

Cryle, P. (2025). *How race theory began to unravel even in its heyday: The (in)discipline of physical anthropology at the time of the Franco-Prussian war (1870–71)* [Manuscript submitted for publication]. Faculty of Humanities, Arts and Social Sciences, The University of Queensland.

Cuomo, F. (2005). *I dieci. Chi erano gli scienziati italiani che firmarono il "Manifesto della razza"* [The ten: Who were the Italian scientists who signed the "Race Manifesto"]. Baldini Castoldi Dalai.

Curran, A. (2012). Buffon et l'histoire naturelle des Africains [Buffon and the natural history of Africans]. *Dix-Huitième Siècle, 44*, 183–199.
Czech, H. (2018). Hans Asperger, national socialism, and "race hygiene" in Nazi-era Vienna. *Molecular Autism, 9*, 1–43.
Dall'Orto, G. (2000). Il paradosso del razzismo fascista verso l'omosessualità [The paradox of fascist racism towards homosexuality]. In A. Burgio (Ed.), *Nel nome della razza. Il razzismo nella storia d'Italia, 1870–1945* [In the name of race: Racism in Italian history, 1870–1945] (pp. 515–528). Il Mulino.
Danziger, K. (1990). *Constructing the subject: Historical origins of psychological research*. Cambridge University Press.
Darwin, C. (1859). *On the origin of species by means of natural selection or the preservation of favored races in the struggle for life*. Murray.
Darwin, C. (1871). *The descent of man and selection in relation to sex* (2 vols.). Murray.
Davidai, S., & Ongis, M. (2019). The politics of zero-sum thinking: The relationship between political ideology and the belief that life is a zero-sum game. *Science Advances, 5*(12), eaay3761.
Davidai, S., & Tepper, S. J. (2023). The psychology of zero-sum beliefs. *Nature Reviews Psychology, 2*(8), 472–482.
de Gobineau, A. (1853–1855). *Essai sur l'inégalité des races humaines* [Essay on the inequality of human races]. Didot.
de Lapouge, G. V. (1899). *L'Aryen, son rôle social* [The Aryan, his social role]. A. Fontemoing.
De Leo, M. (2021). *Queer. Storia culturale della comunità LGBT+* [Queer. A cultural history of the LGBT+ community]. Einaudi.
Della Valle, G. (1911). Le Case dei Bambini e la Pedagogia scientifica di M. Montessori [The Children's Houses and the scientific pedagogy of M. Montessori]. *Rivista Pedagogica, 4*, 67–80.
Di Cesare, D. (2016). *Heidegger e gli ebrei: I «Quaderni neri»* [Heidegger and the Jews: The "Black Notebooks"]. Bollati Boringhieri.
Di Nola, A. M. (1973). *Antisemitismo in Italia 1962/1972* [Antisemitism in Italy 1962/1972]. Vallecchi.
Di Nola, A. M. (1977). Mircea Eliade e l'antisemitismo [Mircea Eliade and antisemitism]. *La Rassegna Mensile di Israel, 43*(1/2), 12–15.
Dreyfus, M. (2011). *L'antisémitisme à gauche. Histoire d'un paradoxe de 1830 à nos jours* [Anti-semitism on the left: The history of a paradox from 1830 to the present day]. La Découverte.
Du Bois, W. E. B. (1939). *Black folk: Then and now: An essay in the history and sociology of the Negro race*. Holt, Rinehart & Winstone.

Duichin, M. (2012). Marx «frenologo». Una pagina rimossa di storia delle idee [Marx "phrenologist": A removed page from history of ideas]. *Intersezioni, 2*, 211–234.

Edelheit, H. (2019). *History of Zionism: A handbook and dictionary*. Routledge

Engels, F. (2002). *Über den Antisemitismus* [About anti-Semitism]. In *Werke, Artikel, Entwürfe, Oktober 1886 bis Februar 1891* [Works, articles, drafts, October 1886 to February 1891] (vol. 31, pp. 1033–1037). Akademie Verlag. (Original work published 1890)

Escudero, J. A. (2015). Heidegger's Black notebooks and the question of anti-semitism. *Gatherings: The Heidegger Circle Annual, 5*, 21–49.

Evola, J. (1942). Il miraggio del Rinascimento [The mirage of the Renaissance]. *Carattere, Rassegna Del Lavoro Italiano, 2*, 5–6.

Evola, J. (1968–2000). *L'arco e la clava* [The bow and the club]. Mediterranee.

Evola, J. (1969). Avvento del «Quinto Stato»? [Advent of the "fifth estate"?]. *Il Borghese, XX*(30), 684–689.

Evola, J. (1997). *Metafisica del sesso* [Metaphysics of sex]. Mediterranee.

Evola, J. (2017). *The myth of the blood: The genesis of racialism*. Arktos. (Original work published 1937)

Evola, J. (2018). *Pagan imperialism and metaphysics of war*. Wewelsburg archives publication. (Original work published 1928)

Evola, J. (2020). *Synthesis of the doctrine of race*. Cariou Publishing. (Original work published 1941)

Eysenck, H. J., & Kamin, L. (1981). *Intelligence: The battle for the mind*. Macmillan.

Faderman, L. (2015). *The gay revolution: The story of the struggle*. Simon & Schuster.

Fanon, F. (1961). *Les damnés de la terre* [The wretched of the earth]. Maspero.

Ferraris, M. (2016). Il complotto di Heidegger [The Heidegger conspiracy]. *Rivista di Estetica, 62*, 133–142.

Ferri, E. (1895). *L'omicidio nell'Antropologia Criminale* [Murder in Criminal Anthropology]. Bocca.

Foschi, R. (2008). Science and culture around the Montessori's first "Children Houses" in Rome (1907–1915). *Journal of the History of the Behavioral Sciences, 44*(3), 238–257.

Foschi, R. (2017). Vino vecchio in botti nuove: I test di intelligenza, la valutazione scolastica e il benessere della società [Old wine in new barrels: Intelligence tests, school assessment, and social well-being]. *International Journal of Psychoanalysis and Education, 9*(1), 47–54.

Foschi, R. (2019). L'eugenetica nella storia della psicologia [Eugenics in the history of psychology]. *Physis. Rivista Internazionale di Storia della Scienza, 1*(2), 291–307.

Foschi, R. (2020). *Storia della psicologia e della mente* [History of psychology and the mind]. Mondadori.
Foschi, R. (2024). "He who leaves the old way for the new, knows what he leaves, but does not know what he will find": Traditionalism, zero-sum thinking, and imagined future. *World Futures, 80*(3), 223–240.
Foschi, R., & Cicciola, E. (2006). Politics and naturalism in the 20th century psychology of Alfred Binet. *History of Psychology, 9*(4), 267–289.
Foschi, R., Giannone, A., & Giuliani, A. (2013). Italian psychology under protection: Agostino Gemelli between Catholicism and fascism. *History of Psychology, 16*(2), 130–144.
Foschi, R., Innamorati, M., & Taradel, R. (2018). «A disease of our time»: The Catholic Church's condemnation and absolution of psychoanalysis (1924–1975). *Journal of the History of the Behavioral Sciences, 54*(2), 85–100.
Foschi, R., & Lauriola, M. (2016). Do amoral familism and political distrust really affect North-South differences in Italy? *Journal of Cross-Cultural Psychology, 47*, 751–764.
Franklin, B. (1763, December 17). From Benjamin Franklin to John Waring. *Founders Online, National Archives*. https://founders.archives.gov/documents/Franklin/01-10-02-0214
Fryer, J. (1985). The unveiling of Dr. Anonymous. *Association of Gay and Lesbian Psychiatrists, 11*(2), 1–4.
Fuchs, C. (2015). Martin Heidegger's anti-semitism: Philosophy of technology and the media in the light of the "Black notebooks". Implications for the reception of Heidegger in media and communication studies. *TripleC: Communication, Capitalism & Critique. Open Access Journal for a Global Sustainable Information Society, 13*(1), 55–78.
Galton, F. (1883). *Inquiries into human faculty and its development*. Macmillan.
Galton, F. (1889). *Natural inheritance*. Macmillan.
Galton, F. (2001). The Eugenic College of Kantsaywhere. *Utopian Studies, 12*(2), 191–209. (Original work published 1911)
Galzerano, G. (1997). *Giovanni Passannante. La vita, l'attentato, il processo, la condanna a morte, la grazia "regale" e gli anni di galera del cuoco lucano che nel 1878 ruppe l'incantesimo monarchico* [Giovanni Passannante: Life, the assassination attempt, the trial, the death sentence, the "royal" pardon and the prison years of the Lucanian cook who in 1878 broke the monarchical spell]. Galzerano Editore.
Genovese, R. (Ed.). (2021). *Sulla sindrome identitaria* [On the identity syndrome]. Rosenberg & Sellier.
Gillham, N. W. (2001). *A life of Sir Francis Galton: From African exploration to the birth of Eugenics*. Oxford University Press.
Gould, S. J. (1987). Petrus Camper's angle. *Natural History, 96*(7), 12–18.
Gould, S. J. (1994). The geometer of race. *Discover Magazine, 15*(11), 64–69.

Gould, S. J. (1996). *The mismeasure of man* (2nd ed.). Norton.
Grant, M. (1916). *The passing of the great race; or, the racial basis of European History*. Charles Scribner's Sons.
Guan, M., & So, J. (2023). Social identity theory. In E. Y. Ho, C. L. Bylund, J. C. M. van Weert, I. Basnyat, N. Bol, & M. Dean (Eds.), *The international encyclopedia of health communication* (pp. 1–5). Wiley.
Guillaumin, C. (1995). *Racism, sexism, power and ideology*. Routledge.
Hacking, I. (1990). *The taming of chance*. Cambridge University Press.
Hacking, I. (2005). *Façonner les gens (II). Cours au Collège de France (2004–2005)* [Shaping People (II). Lectures at the Collège de France (2004–2005)]. https://www.college-de-france.fr/fr/agenda/cours/faconner-les-gens-ii
Hall, L. A. (1990). Illustrations from the Wellcome Institute Library: The Eugenics Society Archives in the Contemporary Medical Archives Centre. *Medical History, 34*(3), 327–333.
Hayden, L. (1866). *Caste among Masons: Address before Prince Hall Grand Lodge of free and accepted Masons of the State of Massachusetts, at the Festival of St. John the Evangelist, December 27, 1865*. Coombs & Co.
Heidegger, M. (2014). *Gesamtausgabe, Band 96: Überlegungen XII-XV* [Complete Edition, Volume 96: Reflections XII-XV]. Klostermann. (Original work published 1939–1941)
Herrnstein, R. J., & Murray, C. (1994). *The Bell Curve: Intelligence and class structure in American Life*. Free Press.
Herzog, D. (2011). *Sexuality in Europe: A twentieth-century history*. Cambridge University Press.
Hilberg, R. (1961). *The destruction of the European Jews*. Quadrangle Books.
Hitler, A. (1940). *My Struggle*. Stalag Edition. (Original work published 1925–1927)
Inglehart, R. (1977). *The silent revolution: Changing values and political styles among Western publics*. Princeton University Press.
Inglehart, R. F., & Norris, P. (2016). *Trump, Brexit, and the rise of populism: Economic have-nots and cultural backlash*. Harvard University, John F. Kennedy School of Government.
Israel, G., & Nastasi, P. (1998). *Scienza e razza nell'Italia fascista* [Science and race in Fascist Italy]. Il Mulino.
Jackson, J. P., & Weidman, N. M. (2006). *Race, racism, and science: Social impact and interaction*. Rutgers University Press.
Jefferson, T. (1787). *Notes on the State of Virginia*. Stockdale.
Jensen, A. (1969). How much can we boost I.Q. and scholastic achievement? *Harvard Educational Review, 39*(1), 1–123.
Jost, J. (2020). *A theory of system justification*. Harvard University Press.

Kanakamedala, P., & Haga, S. B. (2012). Characterization of clinical study populations by race and ethnicity in biomedical literature. *Ethnicity & Disease, 22*(1), 96–101.

Kantrowitz, S. (2010). 'Intended for the better government of man': The political history of African American Freemasonry in the era of emancipation. *The Journal of American History, 96*(4), 1001–1026.

Kevles, D. J. (1985). *In the name of eugenics: Genetics and the uses of human heredity*. University of California Press.

Kiernan, R. (2016). Racism, misogyny and 'motiveless malignity' in Othello. In *Discovering literature: Shakespeare & renaissance*. British Library.

Knox, R. (1862). *The races of men*. Henry Renshaw.

Kuznick, P. J. (2019). *Beyond the laboratory: Scientists as political activists in 1930s America*. University of Chicago Press.

La Vergata, A. (2009). *Colpa di Darwin? Razzismo, eugenetica, guerra e altri mali* [Darwin's fault? Racism, eugenics, war, and other evils]. UTET.

Lachman, G. (2008). *Politics and the occult: The left, the right, and the radically unseen*. Quest Books.

Lachman, G. (2009). *A dark muse: A history of the occult*. Basic Books.

Lachman, G. (2018). *Dark star rising: Magick and power in the age of Trump*. TarcherPerigree.

Lauriola, M., Foschi, R., & Marchegiani, L. (2015). Integrating values and cognitive style in a model of right-wing radicalism. *Personality and Individual Differences, 75*, 147–153.

Le Bon, G. (1879). Recherches anatomiques et mathématiques su les lois des variations du volume du cerveau et sur leurs relations avec l'intelligence. *Revue D'anthropologie, 2*(2), 27–104.

Leahey, T. H. (1980). *A history of psychology: Main currents in psychological thought*. Prentice Hall.

Lewin, K. (1931). The conflict between Aristotelian and Galilean modes of thought in contemporary psychology. *Journal of Genetic Psychology, 5*, 141–177.

Lewontin, R. C. (1972). The apportionment of human diversity. In T. Dobzhansky, M. K. Hecht, & W. C. Steere (Eds.), *Evolutionary biology* (pp. 381–398). Springer.

Lifton, R. J. (1986). *The Nazi doctors: Medical killing and the psychology of genocide*. Basic Books.

Liss, J. E. (1998). Diasporic identities: The science and politics of race in the work of Franz Boas and W. E. B. Du Bois, 1894–1919. *Cultural Anthropology, 13*(2), 127–166.

Lombardo, G. P., & Foschi, R. (2002). *La costruzione scientifica della personalità* [The scientific construction of personality]. Bollati Boringhieri.

Lombroso, C. (1878). *Pensiero e meteore* [Thoughts and meteors]. Dumolard.

Lombroso, C. (1898). *In Calabria*. Giannotta.

Lombroso, C. (1902). L'ultimo brigante. Giuseppe Musolino [The Last Bandit. Giuseppe Musolino]. *Nuova Antologia, 181*, 508–516.

Lombroso, C. (1985). Del parallelismo tra l'omosessualità e la criminalità innata [On the parallelism between homosexuality and innate criminality]. *Sodoma, II*(2), 70–72. (Original work published 1906)

Lombroso, C., & Bianchi, L. (1884). *Misdea e la nuova scuola penale* [Misdea and the new criminal school]. Bocca.

Lombroso, C., & Ferrero, G. (1893). *La donna delinquente, la prostituta e la donna normale* [The female delinquent, the prostitute and the normal woman]. Roux.

Lynn, R. (2010). In Italy, North-South differences in IQ predict differences in income, education, infant mortality, stature, and literacy. *Intelligence, 38*, 93–100.

Lynn, R. (2013). Rushton's r–K life history theory of race differences in penis length and circumference examined in 113 populations. *Personality and Individual Differences, 55*(3), 261–266.

Maiocchi, R. (1999). *Scienza italiana e razzismo fascista* [Italian science and fascist racism]. La Nuova Italia.

Mantegazza, P. (1873). *Fisiologia dell'amore* [Physiology of love]. Bernardoni.

Marx, K. (1844). *On the Jewish question*. https://www.marxists.org/archive/marx/works/1844/jewish-question/

Mayr, E. (1982). *The growth of biological thought: Diversity, evolution, and inheritance*. Harvard University Press.

Meghnagi, D. (2004). A proposito del convegno dell'UCEI tra Vecchio e nuovo antisemitismo [Regarding the UCEI conference on Old and New Antisemitism]. *La Rassegna Mensile di Israel, 70*(3), 185–189.

Milgram, S. (1974). *Obedience to authority: An experimental view*. Harper & Row.

Milicia, M. T. (2014). *Lombroso e il brigante* [Lombroso and the brigand]. Salerno editrice.

Montagu, A. (1952). *Man's most dangerous myth. The fallacy of race. Third edition—Revised and enlarged*. Harper & Brothers.

Montaldo, S. (2019). *Donne delinquenti. Il genere e la nascita della criminologia* [Female delinquents: Gender and the birth of criminology]. Carocci.

Montuori, A. (2005). How to make enemies and influence people: Anatomy of the anti-pluralist, totalitarian mindset. *Futures, 37*(1), 18–38.

Moore, C. (1974). Were Marx and Engels white racists? The prolet-aryan outlook of Marxism. *Berkeley Journal of Sociology, 19*, 125–156.

Moretti, E. (2021). *The best weapon for peace: Maria Montessori, education, and children's rights*. University of Wisconsin Press.

Mosse, G. L. (1978). *Toward the final solution: A history of European Racism.* Fertig.

Myrdal, G. (1944). *An American dilemma: The negro problem and modern democracy.* Harper & Bros.

Nash, G. B. (2006). Franklin and slavery. *Proceedings of the American Philosophical Society, 150*(4), 618–635.

Niceforo, A. (1898). *L'Italia barbara contemporanea* [Contemporary barbaric Italy]. Sandron.

Niceforo, A. (1901). *Italiani del Nord e Italiani del Sud* [Northern Italians and Southern Italians]. Bocca.

Pende, N. (1933). *Bonifica umana razionale e biologia politica* [Rational human reclamation and political biology]. Cappelli.

Pernick, M. S. (1996). *The Black stork.* Oxford University Press.

Piasere, L. (2021). *Antisemitismo e antiziganismo in Italia. Sull'antropologia del razzismo di Alfonso Maria di Nola* [Antisemitism and antiziganism in Italy: On the anthropology of racism of Alfonso Maria di Nola]. Morlacchi.

Pick, D. (1989). *Faces of degeneration: A European disorder, c. 1848–1918.* Cambridge University Press.

Poidimani, N. (2009). *Difendere la razza. Identità razziale e politiche sessuali nel progetto imperiale di Mussolini* [Defending the race: Racial identity and sexual policies in Mussolini's imperial project]. Sensibili alle foglie.

Prophet, E. (2024). *Racist or liberating? The difficult transformations of theosophical root race theory.* In C. M. Stang & J. A. Josephson Storm (Eds.), *Theosophy and the study of religion* (pp. 145–173). Brill.

Protasi, M. R., & Sonnino, E. (2003). Politiche di popolamento: Colonizzazione interna e colonizzazione demografica nell'Italia liberale e fascista [Settlement policies: Internal colonization and demographic colonization in liberal and fascist Italy]. *Popolazione e Storia, 4*(1), 91–138.

Proudhon, P.-J. (1847). *On the Jews.* https://www.marxists.org/reference/subject/economics/proudhon/1847/jews.htm

Putnam, R. D. (1993). *Making democracy work.* Princeton University Press.

Ray, L. J. (1983). Eugenics, mental deficiency and Fabian socialism between the wars. *Oxford Review of Education, 9*(3), 213–222.

Richards, G. (1997). *'Race', racism and psychology: Towards a reflexive history.* Routledge.

Roberts, R., & Davidai, S. (2022). The psychology of asymmetric zero-sum beliefs. *Journal of Personality and Social Psychology, 123*(3), 559–575.

Romeo, D., & Cefalì, F. (2015). *Salvatore Misdea. 1884: follia criminale o determinazione di un soldato del sud Italia?* [Salvatore Misdea. 1884: Criminal madness or determination of a southern Italian soldier?]. Gigliotti.

Rose, N. (1985). *The psychological complex: Psychology, politics and society in England, 1869–1939.* Routledge & Kegan Paul.

Rose, N. (2007). *The politics of life itself: Biomedicine, power, and subjectivity in the twenty-first century*. Princeton University Press.

Rossi, P. (1898). *L'animo della folla* [The soul of the crowd]. Riccio.

Rota, G. (2003). Un filosofo razzista. Note su Evola [A racist philosopher. Notes on Evola]. *Rivista di Storia della Filosofia*, *58*(3), 459–496.

Różycka-Tran, B., Boski, P., & Wojciszke, B. (2015). Belief in a zero-sum game as a social axiom: A 37-nation study. *Journal of Cross-Cultural Psychology*, *46*(4), 525–554.

Said, E. (1978). *Orientalism*. Pantheon Books.

Saini, A. (2017). *Inferior: How science got women wrong—and the new research that's rewriting the story*. Beacon Press.

Saini, A. (2019). *Superior: The return of race science*. Beacon Press.

Sale, G. (2002). Antigiudaismo o antisemitismo? Le accuse contro la chiesa e la «Civiltà Cattolica» [Anti-Judaism or antisemitism? Accusations against the church and Civiltà Cattolica]. *Civiltà Cattolica, Quaderno*, 3647(II), 419–431.

Salvemini, G. (1900). *Risposta* [Reply]. In A. Renda (Ed.), *La questione meridionale. Inchiesta* [The southern question. Inquiry] (pp. 38–42). Sandron.

Salvemini, G. (1955). *Scritti sulla questione meridionale (1896–1955)* [Writings on the Southern Question (1896–1955)]. Einaudi.

Saretzky, G. D. (1982). *Carl Campbell Brigham, the native intelligence hypothesis, and the Scholastic Aptitude Test*. Educational Testing Service, Research Publications R-116.

Schirripa, P. (2021). L'urgenza di un'antropologia dell'antisemitismo [The urgency of an anthropology of antisemitism]. In L. Piasere (Ed.), *Antisemitismo e antiziganismo in Italia. Sull'antropologia del razzismo di Alfonso Maria di Nola* [Antisemitism and antiziganism in Italy. On the anthropology of Alfonso Maria di Nola's racism] (pp. 107–113). Morlacchi.

Sergi, G. (1916a). La cultura Germanica [The Germanic culture]. *Nuovo Convito*, *1*, 7–9.

Sergi, G. (1916b). Gli istinti gregari nell'uomo e il popolo tedesco [The gregarious instincts in man and the German people]. *Nuovo Convito*, *1*, 151–154.

Sheffer, E. (2018). *Asperger's children: The origins of autism in Nazi Vienna*. Norton.

Sidanius, J., Pratto, F., & Devereux, E. (1992). A comparison of symbolic racism theory and social dominance theory as explanations for racial policy attitudes. *The Journal of Social Psychology*, *132*(3), 377–395.

Sighele, S. (1891). *La folla delinquente* [The delinquent crowd]. Fratelli Bocca.

Spektorowski, A., & Mizrachi, E. (2004). Eugenics and the welfare state in Sweden: The politics of social margins and the idea of a productive society. *Journal of Contemporary History*, *39*(3), 333–352.

Stanton, E. C. (1889). *A history of woman suffrage* (Vol. 1). Fowler & Wells.

Taguieff, P.-A. (1994). *La forza del pregiudizio. Saggio sul razzismo e sull'antirazzismo* [The force of prejudice: An essay on racism and anti-racism]. Il Mulino. (Original work published 1988)

Taguieff, P.-A. (2005). Racisme aryaniste, socialisme et eugénisme chez Georges Vacher de Lapouge (1854–1936) [Aryanist racism, socialism and eugenics in Georges Vacher de Lapouge (1854–1936)]. *Revue D'histoire De La Shoah, 183*(2), 69–134.

Taguieff, P.-A. (2011, July 15). *L'antiracisme, une machine à exclure?* [Antiracism, a machine of exclusion?]. Atlantico. https://atlantico.fr/article/decryptage/antiracisme-sos-racisme-concert-pour-egalite-14-juillet

Tajfel, H. (1981). *Human groups & social categories*. Cambridge University Press.

Taradel, R. (2002). *L'accusa del sangue: storia politica di un mito antisemita* [The blood accusation: Political history of an anti-Semitic myth]. Edizioni Riunite.

Tarquini, A. (2019). *La sinistra italiana e gli ebrei. Socialismo, sionismo e antisemitismo dal 1892 al 1992* [The Italian left and the Jews: Socialism, Zionism, and anti-Semitism from 1892 to 1992]. Il Mulino.

Teti, V. (2011). *La razza maledetta. Origini del pregiudizio antimeridionale* [The cursed race: Origins of anti-Southern prejudice] (2nd ed.). Manifestolibri.

Topinard, P. (1888). Le poids de l'encéphale d'après les registres de Paul Broca [Le poids de l'encéphale d'après les registres de Paul Broca]. *Mémoires Société d'Anthropologie Paris, 2nd series, 3*, 1–41.

Trawny, P. (2015). *Heidegger and the myth of a Jewish world conspiracy*. University of Chicago Press.

Tucker, W. H. (2007). *The Funding of Scientific Racism: Wickliffe Draper and the Pioneer Fund*. University of Illinois Press.

Tucker, W. H. (2009). *The Cattell Controversy: Race, Science, and Ideology*. University of Illinois Press.

UNESCO. (1950). *The race question*. UNESCO Publication 791. https://unesdoc.unesco.org/ark:/48223/pf0000128291

UNESCO. (1969). *Four statements on the race question*. UNESCO. https://unesdoc.unesco.org/ark:/48223/pf0000122962

UNESCO. (1979). *Declaration on race and racial prejudice adopted by the General Conference of UNESCO at its twentieth session Paris, 27 November 1978*. UNESCO. https://unesdoc.unesco.org/ark:/48223/pf0000039429

Vaksberg, A. (1994). *Stalin against the Jews*. Knopf.

Van Ginneken, J. (1992). *Crowds, psychology, and politics, 1871–1899*. Cambridge University Press.

Vincenzo, C., & Foschi, R. (2021). *Razzismo vecchio stile, razzismo moderno, populismo: una ricerca di psicologia sociale* [Old style racism, modern racism, populism: A social psychology study]. In R. Genovese (Ed.), *Sulla sindrome identitaria* [On the identity syndrome] (pp. 23–70). Rosenberg & Sellier.

Walsh, R. T., Teo, T., & Baydala, A. (2014). *A critical history and philosophy of psychology: Diversity of context, thought, and practice*. Cambridge University Press.

West, C. (1982). *Prophesy deliverance! An Afro-American revolutionary Christianity*. The Westminster Press.

Winston, A. S. (2018). Neoliberalism and IQ: Naturalizing economic and racial inequality. *Theory & Psychology, 28*(5), 600–618.

Yudell, M., Roberts, D., DeSalle, R., & Tishkoff, S. (2016). Taking race out of human genetics. *Science, 351*(6273), 564–565.

Zanardelli, G. (1881). *Discussioni del 4 aprile 1881* [Discussions of April 4, 1881]. In *Atti del Parlamento italiano - Discussioni della Camera dei Deputati, XIV Legislatura - Sessione 1880 - 1881 (24/02/1880 - 09/04/1881), Volume V dal 24/02/1880 al 09/04/1881* [Proceedings of the Italian Parliament—Debates of the Chamber of Deputies, XIV Legislature—Session 1880—1881 (February 24, 1880–April 9, 1881), Volume V from February 24, 1880 to April 9, 1881] (pp. 5001–5016). Tipografia Eredi Botta.

Zenderland, L. (2001). *Measuring minds: Henry Herbert Goddard and the origins of American intelligence testing*. Cambridge University Press.

Zimbardo, P. G. (2007). *The Lucifer effect*. Random House.

Index

R. Foschi, *Racisms*, https://doi.org/10.1007/978-3-032-22962-5

The manufacturer's authorised representative in the EU is Springer Nature Customer Service Centre GmbH, Europaplatz 3, 69115 Heidelberg, Germany. If you have any concerns regarding our products, please contact ProductSafety@springernature.com

Printed and bound by CPI Group (UK) Ltd, Croydon, CR0 4YY
17/09/2026
02218454-0003